CALIFORNIA

VANISHING HABITATS AND WILDLIFE

CALIFORNIA

VANISHING HABITATS AND WILDLIFE

Text and Photography by

B. "Moose" Peterson

Foreword by

Roger Tory Peterson

Beautiful America Publishing Company
T.M.

Joshua Tree. *On the highest perch it can find in a Joshua tree, a black-throated sparrow sings to attract a mate and announce his territory.*

This book is dedicated to the wildlife of California,

which has touched my heart and

become part of my spirit.

Beautiful America Publishing Company©
P.O. Box 646
Wilsonvillc, Oregon 97070

Library of Congress Cataloging-in-Publication Data
Peterson, B. "Moose"
California, vanishing habitats & wildlife / text and
photography by B. "Moose" Peterson
p. cm.
ISBN 0-89802-589-3
1. Natural history —California. 2. Rare animals—
California. 3. Endangered species— California. I. Title.
II. Title: California, vanishing habitats and wildlife.
QH105.C2P52 1993
508.794—dc20 92-47262
CIP

Design: Jacelen Pete
Editing: Andrea Tronslin
Linotronic output: LeFont Laser Imaging & Design
Printed in Korea

TABLE OF CONTENTS

ACKNOWLEDGMENTS

To my field crew: my wife Sharon and my sons Brent and Jake. To the many biologists who have generously given me their time, as well as an education unmatched anywhere: Meaton Freel, Dan and Sue Williams, Charlie Collins, Paul Kelly, Linda Spiegel, Vern Bleich, Paul Collins, Steve Laymon, Steve Montgomery, Dick Zembal, Dick Anderson, Bob and Pat Brown-Berry, Bill LaHaye, Esther Burkett, Dan Christianson, Ken Sasaki, Ron Thomas, Gary Gulasi, Dave Clendenen, Jim Estep, Wayne Ferren, Denise LaBerteaux, Philip and Barbara Leitner, June Mire, Peter Moyle, Richard Smith, Kevin Foerster, Ron Powell, Ted Murphy, Gary Bell, Cameron Barrows, Gary Strachen, and Karen Kakiba-Russell.

Thanks to those who were a tremendous aid in my research for all my projects: Ron Schlorff, John Gustafson, Mike Bender, Gordon Gould, Roy Woodward, Gail Kobetich, Shirley Clarke, Terri Sheridan, and a special thanks to Dave Dick who published my first article and photographs, and who has been a good friend and supporter from the start.

LEFT: ***Tule Lake,*** *mule deer.*

Foreword

Roger Tory Peterson

To most Americans as well as many people around the world, California is the Promised Land. It has more diversity within its boundaries than any other state in the lower forty-eight with the possible exception of Texas. But even Texans admit to a yearning for California.

Because of California's allure, people have poured in by the millions. During the gold rush days of the 1840s and 1850s, many pioneers came primarily to seek reward in the mines. During the 1930s, drought forced farmers in the dust bowl to seek survival in the valleys further west. But today, when settlers come from the East or Midwest, they simply want to escape the winter and the mud season by taking residence in the Sunshine State.

When I first visited California some fifty years ago while working on my *Field Guide to Western Birds*, there were about seven million residents in the state. Some areas were already overcrowded. Now there are almost thirty million and the flow continues. Because of this influx, some of the older residents are now moving out, moving north into Oregon, Washington, Idaho, even Montana and British Columbia.

All this pressure by the world's ever-increasing population is having a disastrous effect on the environment and wildlife. Man, the dominant primate, holds the whip hand over every other creature on earth. But there are those among us who believe it is our moral duty, the humane thing, to defend wildlife as best we can. We should not be so egocentric as to believe that only we, as humans, should survive.

Moose Peterson (no relation of mine), who is still a young man, has dedicated his life to the California ethic. His medium of enlightenment is the camera; no one is more skilled as a wildlife photographer. Not only is he well-connected with Nikon and Canon, but he also conducts photographic workshops to Mono Lake, Point Reyes, and many other ecological hot spots in the state. He has managed to record on film most of the threatened species on the California endangered list, not only the birds and mammals, but also reptiles, fish, and even some of the insects. He has amassed a file of photographs of his home state unequalled by any of his peers.

Consider the biodiversity of California, flanked on the West by eight hundred miles of Pacific coast with its beaches, rocky headlands, and offshore islands, and on the East by the Sierra Nevada with its snow-covered peaks. The humid coastal ranges and the high Sierra are separated by the broad central valleys, the Sacramento and the San Joaquin. The dry deserts of the far Southeast are in sharp contrast to the humid forests of the Northwest. In a drive of a couple of hours or less, you can go from surfing to skiing. It is easy to see why residents of Nebraska or Iowa would crave more than the occasional visit; after a taste, they might even decide to retire there.

Although several of our largest and oldest national parks are in California, some like Yosemite are literally loved to death. Tourism is very important in saving the environment — "ecotourism" it has been called — yet there are limits to the number of visitors that a publicized beauty spot like the Yosemite Valley can accommodate.

Except for the house sparrow, starling, and domestic pigeon, long habituated to the urban world, and those few native birds that have learned the tricks of living with us, wildlife is in trouble. Consider beach birds such as the least tern and the snowy plover, local and endangered in California. Both are especially vulnerable because of sunbathers, picnickers, beach buggies, and other vehicles on their limited nesting grounds. Only because of the vigilance of a few dedicated people, who string wires around the nesting sites and stand guard, do we still have these gentle birds.

Because birds with their sharp eyes are observant, some have learned that our wasteful ways can be turned to their advantage. Crows know that they will not be shot when they furtively investigate the trash cans behind a fast food restaurant. Gulls are more open about it, and when soliciting food may become as trusting as pigeons. Only dogs that chase them and kids under four who run at them with flailing arms are to be avoided.

Wetlands, so important to wildlife, are under constant stress because of agriculture, drainage, mosquito control, and toxic waste.

Everyone knows about the California condor which in recent times decreased from a population of hundreds to about two dozen individuals. These were finally trapped and put in the custody of zoos in Los Angeles and San Diego, where their chicks are being raised. After introducing Andean condors to act as role models for these young birds, they are now being returned to the wild. We may yet save this giant Californian which was so close to the brink.

Nowhere in America are the battle lines more sharply drawn than in California between those who want to preserve and those who would exploit. Moose Peterson understands the problems as well as anyone, certainly better than I whose home is on the East Coast, 3,400 miles away. However, as a world traveller, I can put things into perspective; I will leave it to Moose to give you the details about his beloved state. He is both an historian and an elucidator through his photography, a teacher, and an opinion maker. He makes us aware. Awareness leads to concern and concern leads to action.

Don't simply skim the pictures in this book. Read what Moose has to say and think about it.

GRASSLANDS

The California Prairie

We camped under the trees, in a ravine among the hills, from which we could look down over the plains below, and see the grassy undulations rolling gently away as far as the eye could reach. It was a beautiful spot, destined some day to present as splendid a domestic picture as now a wild one; and, as my eyes dwelt delighted upon it, I could not help thinking of a time when millions of stock shall crop the herbage there, and the now wild landscape be studded with farms and villages of a numberless and thriving population. This will only be, thought I, when these old bones of mine shall have crumbled into dust, and when the memory of the present inhabitants shall be preserved only in old stories.

James Capen Adams (Grizzly Adams) 1853

The "historic" descriptions of California's Central Valley, such as that of Grizzly Adams, seem almost fanciful. While speeding along on Interstate 5 today, looking at the landscape, it is hard to conjure up images of vast inland lakes or herds of pronghorn prancing through the grasslands, or the possibility of coming face to face with a California grizzly bear crossing the highway. It is even harder to comprehend that we have managed to dismantle nearly all visual record of the landscape that filled the Central Valley in just a hundred years!

LEFT: ***Elkhorn Plains.*** *The setting sun lights the dry arroyo and the shrubs that line its banks. These arroyos can be runways for San Joaquin kit foxes coming down into the plains, as well as homes for other small mammals pushed out by the giant kangaroo rat.*

The San Joaquin Valley is about 250 miles long and forty to sixty miles wide. It originally comprised 8.5 million acres of native habitat, but now only three percent remains—approximately 370,000 acres.

The valley floor is not level, but slopes downhill from south to north. Starting at an elevation of three hundred feet near Grapevine, it drops to sea level at the delta west of Sacramento. This, coupled with being surrounded by mountain ranges on three sides, has made the grassland prairie region unique.

"Islands" of different habitats within the valley are determined mostly by the water they receive. Rainfall of five to six inches in the south and fourteen inches in the north is just part of the water equation. The other factor is runoff from the many rivers and streams that flow from the bordering mountains, making many regions within the valley susceptible to annual flooding. The vast quantity of free-flowing water also gives life to fresh water marshes, lakes, and riparian forests.

The valley supported the largest lake west of the Mississippi a hundred years ago. Three large, interconnected lakes were fed by runoff but had no outflow. Tulare Lake, the largest of the three, had a surface of thirty-three by twenty-one miles. These lakes and the fresh water marshes at their edges were surrounded by vast acres of tall reeds called tules, which were the home for a magnificent animal, the tule elk.

The tule elk spent much of their time grazing in the tules. From old accounts, the scene of the large lakes with herds of elk and pronghorns sounds incredibly idyllic. Circling these herds were California grizzly bears and gray wolves, looking for the opportunity for a meal. By 1920, this scene had all but disappeared. The grizzly can be seen only on the state flag, the wolf in old black and white photos; the lakes are dried plots of land. The remaining twenty-eight tule elk were given protection on a ranch near Bakersfield, and have made a slow but steady comeback, spawning a number of herds that live today throughout the state.

The grasslands and the critters that call them home have not fared any better. The California prairie is a hostile community; its niche in the valley is where heat and lack of water are prevalent. Summertime temperatures are near the century mark, winter is in the low thirties, and time for plant growth is short. Despite this, the wildlife that has evolved to live in this habitat does so very successfully, as long as the native habitat remains intact.

Carrizo and Elkhorn Plains

In the spring, native bunch grasses, reaching as tall as the side of a horse, grew thick on the undulating land turning to naturally cured hay in the summer. Wild horses, elk, deer, and antelope were abundant in the plain and large flocks of sandhill cranes spent each winter at Soda Lake.

San Luis Obispo Tribune, 1886

The majesty of California's grassland can still be witnessed. The 82,000-acre Carrizo and Elkhorn plains west of Bakersfield provide a window to a world largely gone. Civilization has passed by its seemingly uninviting landscape nestled in the lower southwestern corner of the Central Valley.

The birth of the Carrizo and Elkhorn plains is an interesting story in itself. The mighty San Andreas earthquake fault runs through the Temblor Mountains on the plains' eastern border. Over thirty million years ago it shook, causing the land to subside and a basin to form. Eons of rain eroded away the mountains, filling in the basin and creating a plain. With no outlet for the water other than evaporation, the soil's alkali was concentrated in what is now Soda Lake. The same alkali influences the very nature of the soil.

Many regions of the San Joaquin Valley were alkali habitats, but they have vanished. The alkali sink, alkali wetlands, saltbush scrub, and annual grasslands that they supported went with them. On the Carrizo, examples of all these habitats still remain intact. They begin in and around Soda Lake, with changes in habitat zones occurring at increasing elevation and distance from the lake.

Farming and a long legacy of cattle ranching, dating back more than a century, disrupted the balance of life on the plains. Native grasses were replaced by exotics and rodenticides were used to control "ground" pests, wiping out mostly non-targeted species. Despite this tampering, the tenacious native wildlife held on. Today as the plains celebrate liberation from the plow and cattle ranching, native wildlife is working towards the balance of long ago.

The environment of the plains is a harsh one—little more than a modified desert habitat. As in any desert, the visual cycle of life begins with the coming of the winter rains.

As if directly pouring out color from the clouds, the rain sets the plains ablaze. Carpets of wildflowers rush to meet the horizon in a celebration of the birth of spring. Yellows, purples, reds, and whites mix and fold, swirl and dance in the spring breeze. The one color missing is the orange of the California poppy, which is absent most years on the plains.

Many of these now endangered wildflowers are indigenous to the Carrizos and Elkhorn. The California jewelflower was once thought to have vanished from the Carrizo, but a small population was rediscovered in the mid-1980s. The blossoms of this two-foot-tall plant glow in the sunlight like jewels strung together by a fine, green thread. The jewelflower resides in an arroyo on the Carrizo, up the slope from the valley floor. Its special home commands a special view of the lower Carrizo Plain.

The San Joaquin woolly-thread is what is commonly referred to as a "belly rubber." That's because you have to get down on your belly to see it. Standing merely two to three inches high, it easily can be overlooked while you are walking the plains. It too has a special niche, in an arroyo on the valley floor where runoff from spring rains is very slight.

The native grasses themselves are rare.

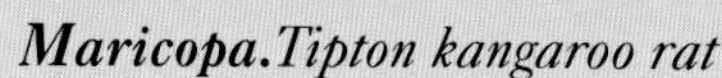

***Maricopa.** Tipton kangaroo rat*

***Turlock.** Fresno kangaroo rat*

Kangaroo Rats

The most-impacted and least-known species squeezed out by agricultural conversion is the kangaroo rat. These industrious little rodents are the natural farmers of the grasslands. Their activities support not only their own existence, but also that of other wildlife.

The San Joaquin kangaroo rat inhabits most of the valley. It is separated into three subspecies: the short-nosed, Tipton, and Fresno kangaroo rats. Of these three, the Tipton and Fresno are federally listed as endangered; the short-nosed probably should be. The subspecies are distinguished not by appearance but by where they are found in the valley.

The Tipton kangaroo rat inhabits the drier southwestern portion of the valley, but only an estimated 3.7 percent of its historic range. Its busy nocturnal activities focus on the harvesting and planting of seeds. These activities, in combination with its burrows, make it a keystone species. It is thought to be facing extinction by the year 2000 because its population has been fragmented into six small groups, leading to inbreeding and possibly its final demise.

The Fresno kangaroo rat lives further north, west of Fresno. Its existence has been precarious since its first discovery. Soon after it was discovered in 1891, it was thought to have become extinct because of habitat conversion to cropland. It was "rediscovered" in 1934 and found to inhabit a slightly greater range than originally believed. Today, efforts to trap a Fresno kangaroo rat have been fruitless and its future remains in grave doubt. The Fresno is unique in that it inhabits wetter regions, especially alkali sinks bordering on waterways, where damp soils and annual flooding make seed storage difficult. Because of this, the Fresno eats more green plant material and insects than the other kangaroo rats.

The bluegrass is a perennial bunchgrass which lost its hold after the introduction of exotics. Grasses such as red brome and red-stem filaree were deemed better feed for cattle and easy to grow. These exotic grasses now dominate much of our native grasslands. Finding native grasses can be as challenging as finding some wildflowers!

One of the most important residents of the plains is the giant kangaroo rat. Like so many of the state's kangaroo rats, it is a keystone species endangered with extinction. Reduced to living on just two to three percent of its historic range, it has found a seemingly secure home on the Carrizo and Elkhorn plains.

Giant kangaroo rats are nothing like their late night horror show cousins. Fourteen inches long, with a furred tail, large hind legs, external fur-lined cheek pouches, and big ears, they are uniquely adapted. They use their large hind legs and sharp nails to dig burrows five to six inches below the surface, constructing large systems. They are the largest of the kangaroo rat species, and in their habitat are very successful at excluding other small rodents. But it is their nighttime activities that make them the cornerstone of the plains, maintaining its habitat ecosystem.

With the rains come the grasses, and life for the giant kangaroo rats becomes the busiest. These energetic creatures spend much of the night harvesting the fresh new seed pods of the grasses covering their precincts. A precinct is the home for one giant kangaroo rat and includes the ground above its burrow system (many precincts make up a colony). Once the seed pods are cut, they are collected and placed in a pile, creating a miniature hay stack. Here the seeds dry in the warm sun before the giant kangaroo rat takes them underground for safe storage.

The giant kangaroo rat also collects individual seeds it finds during its nighttime

Caliente Mountains, Carrizo Plains.
This is morning flight, the time the flocks of lesser sandhill cranes take to the air from their night roost at Soda Lake. They divide their time here between foraging in the fields and roosting in the safety of Soda Lake.

Soda Lake, Carrizo Plains. *The alkali marsh of the Carrizo Plains is one of the few remaining examples left in California. A seasonal marsh, it is dependent on the winter rains to fill, and lasts only as long as the summer sun permits. Runoff from the Caliente Mountains in the background contributes to the water level of Soda Lake and the marsh.*

Soda Lake, Carrizo Plains. *The lesser sandhill cranes celebrate their winter haven. Heading out in the morning to feed, they'll return to Soda Lake around mid-day and then again at dusk.*

rounds. The seeds are cached in small pits around the entrance of its burrow. And like so many species that cache seeds, the giant kangaroo rat does not always collect all it has hidden away. So the combined activities of the giant kangaroo rat—burrow digger, harvester, and seed planter—have a tremendous influence on the flora of the plains. This would seem enough for the life of one small critter, but their importance to the ecosystem goes even further. Their burrows, which provide the giant kangaroo rat a home, an escape from the summer heat, and storage for their seed stores, also provide a home for the blunt-nosed leopard lizard.

Another of the endangered species that lives on the plains, the blunt-nosed leopard lizard is a stunning reptile. Its fourteen inches of length does not provide it with any extra courage. Like so many lizards, it scurries off at the first sign of approaching humans. It enjoys the warm summer temperatures, but not the winter chill, so it finds refuge in the giant kangaroo rat's burrow. Though this relationship is not totally understood, it seems to be a long-established one. The lizard not only finds shelter in the burrow, but also raises its family in its secure depths. The adults emerge in early spring while the young see the plains for the first time in early summer.

Does it sound like the burrows of the giant kangaroo rat are a busy place? There's more. The San Joaquin antelope squirrel, a threatened species, shares the burrow as well. This relatively small squirrel does not dig its own burrow in the hard soils of the plains, so, like the lizard, it depends on the giant kangaroo rat's burrow. Its dependency on the giant kangaroo rat goes further, since the squirrel also eats the seeds produced by the giant kangaroo rat's activities. This relationship of the three species is fascinating to watch unfold each day on the plains, especially through the changes of

LEFT: ***Carrizo Plains.*** *California jewelflower.*

RIGHT: ***Elkhorn Plains.*** *Barely larger than a fist, the threatened San Joaquin antelope squirrel tackles life in a hostile environment. Antelope squirrels curl their long tails over their backs to provide themselves a little shade from the hot sun.*

the seasons. But the role of the giant kangaroo rat does not stop here; it serves one other endangered species.

The San Joaquin kit fox truly represents the wilds of California's grasslands. The kit fox is the smallest fox species in North America, and the San Joaquin is the largest of the four or five subspecies. The size of a small house cat, the fox depends on small prey. On the Carrizo and Elkhorn plains, the giant kangaroo rat serves its last role as a food source for the fox.

The San Joaquin kit fox can coexist with humans and their activities, but only when it is left its own habitat to forage and range. The foxes have been known to den one mile deep in an active oil field, where the parents made constant, vigilant trips outside the field to forage for food to feed their young. But these are the exceptions, since most of man's activities are very unforgiving. The kit fox needs lots of space and with such severe losses to its habitat, it is in desperate trouble.

The kit fox is constantly on the move. Within its home range, it may have as many as seventy different den sites. Why does it move so much? Its only threat besides man is the coyote, who might prey on the fox to eliminate a possible competitor as often as to find a food source. One of the paradoxes of the ecosystem is that the coyote might be a champion of the kit fox.

The introduced (non-native) red fox is moving in on the kit fox, not only taking its food, but actually killing it. The coyote is

the only natural predator of the red fox. With the coyote in the ecosystem, the red fox is eliminated and the kit fox is protected a little longer, at least for most of its range.

San Joaquin kit foxes usually settle down in their natal dens in the spring and produce a litter of three to five small pups. The female stays with them in the den and the male forages for food. To avoid detection by coyotes, the foxes may move to a new den within a day of the pups' birth.

The pups, pudgy, fluffy, and cute, stay

Caliente Mountains, Carrizo Plains. *No more than a few hundred feet above the floor of the plains, the landscape changes dramatically. The plants and animals that call this region home differ from those below. This is the special habitat of the California jewelflower.*

in the safety of the den for the first few weeks. But they can only stand being penned up so long and soon take all their energy to the surface to play in the sun. They spend the next few months with their parents as they grow and learn the ways of the grasslands. The odds of all the litter making it through the first year are very slim. But that's the way it should be; if the population were healthy, there would be no problems. But outside forces—the red fox and feral dogs—are putting additional pressure on the already endangered San Joaquin kit fox, pushing the fox towards extinction.

More than mammals inhabit the grasslands. Every winter, three to five thousand lesser sandhill cranes spend the winter on Soda Lake. These majestic, gray messengers of life roost in the shallow waters on the north end of Soda Lake (when it has water) and spend their days foraging in the fields. They have long been the main attraction for many people who come to the plains. During their typical visitation, from December until March, they fill the sky and fields with grace and song.

But they are not the only celebrated aerialists of the plains. California condors once spent some of the winter soaring over the plains. This prehistoric remnant disappeared from the skies for five years when all the condors were taken into captivity in the 1980s. Old stories of them feeding in large flocks on a carcass or nesting locally seem unbelievable today. One of the brighter notes for the plains is that two birds were released back into the wild on January 14, 1992. Someday we might again witness these majestic birds flying the skies of the Carrizo Plain.

During the winter, the skies of the Carrizo are abuzz with birds. Because of the large rodent population and basically mild climate, the Carrizo Plain is a extensive wintering site for raptors. Bald eagle, golden eagle, osprey, peregrine falcon, prairie falcon, red-tailed and dark phased red-tailed, ferruginous, rough-legged, and Swainson's hawk are just part of the gathering. The birds and wildlife seem to come to the Carrizo to celebrate; watching them it is hard to imagine they are only a small sampling of what once inhabited California's prairies.

Carrizo Plains. *San Joaquin woolly-thread.*

Carrizo Plains. *Gold fields carpet the plains in a wash of gold, easily explaining how this miniature wildflower got its name. The rains which broke the drought unleashed a flood of color, as wildflowers of every size, shape, and color covered the ground.*

LoKern

LoKern could be called the Carrizo/Elkhorn's sister in that they share much of the same habitat. Butting up against the Elkhorn Hills west of Buttonwillow, the four thousand acres of the LoKern are a small fraction of its once greater whole. Though the Temblor Mountains divide the Carrizo/Elkhorn from the LoKern, many species live in both areas. The one feature that does distinguish them is that LoKern is located within the San Joaquin Valley, tucked away on its southwestern edge. Because of this, LoKern protects a couple of other native endangered species not found over the hill.

The LoKern is the last significant intact remnant of native San Joaquin saltbush and alkali sink scrub habitat. A major plant of the grasslands is the alkali-tolerant saltbush. It in turn supports other smaller plants that take advantage of its shade and of the nutrients the dead plant material adds to the soil. The endangered California jewelflower, Hoover's woolly-star, and Kern mallow all reside in this plant community. In fact, the entire world population of Kern mallow lies in the safety of LoKern.

LoKern. *A San Joaquin kit fox family, four pups and an adult, enjoy their time out in the sun. They are intent on watching the female coming back, hoping she was successful in finding a meal. Life for them is simple now, but will get harder as spring turns into summer.*

The saltbush also provide homes and cover for many bird species; the most notable is the LeConte's thrasher. This brownish bird, robin-sized, is not very showy, nor does it do anything really special. But like so many birds and animals, much of its specialized, desert-grassland habitat has been lost, and LoKern is the only stronghold left to it in the valley.

This diverse habitat also supports a number of birds under consideration for listing as endangered—the mountain plover, long-billed curlew, and ferruginous hawk. They are winter visitors to LoKern, depending on its unique features to supply them with the needed food reserves for their return migration north.

This rich plant life also supports the short-nosed, Tipton and giant kangaroo rat, San Joaquin pocket mouse, blunt-nosed leopard lizard, and San Joaquin antelope squirrel. These in turn support the San Joaquin kit fox. The web of life in this "desert" community staggers the imagination. And LoKern supports some of the largest known concentrations of these endangered wildlife and plants, an island unto itself.

Yosemite Valley. *Listed as a predator of the San Joaquin antelope squirrel, giant kangaroo rat and even the San Joaquin kit fox, the coyote might end up being champion to all. This cagey native is the main predator of the exotic red fox, and the fox is a threat to many of the same endangered species.*

ABOVE: ***Elkhorn Plains.*** *The drought severely impacted some populations of giant kangaroo rat because of the lack of plant production. This bumper crop of plants means a bumper crop of baby kangaroo rats; for some populations it is just in the nick of time.*

ABOVE RIGHT: ***Elkhorn Plains.*** *This aerial view of the giant kangaroo rat's home, a precinct, shows its nighttime activities. The trails mark where the rat has mowed down the grasses while harvesting seeds. The pile of seeds is stacked to dry in the summer sun before being stored underground.*

The drama and dynamics of the Carrizo/Elkhorn plains and LoKern does not mirror all of California's grasslands. Each location in the valley is host to a varying group of the same basic species, those that are endangered as well as those that are not. But the interrelationship of species on the Carrizo and Elkhorn plains best depicts what is at stake when one species is lost from the ecosystem.

What of the Carrizo, Elkhorn, and LoKern themselves? The Carrizo and Elkhorn plains are protected now. In the future, more land may be acquired to link them and the condor sanctuary, creating the nation's first macropreserve. LoKern is slowly being purchased, its lands secured from future development. These lands are not fenced off to strangers, but rather await, with an open invitation, anyone seeking to explore these magical places. Take a step back into time, and see California as our ancestors first saw and loved it. The inland sea of grasses waving

in the Pacific breeze is California's unique prairie.

Neither these lands nor their inhabitants are completely safe, but many individuals and agencies have made great commitments to protect them. The healthy scheme of life on the plains is a tribute to the tenacity of its wildlife to hold on, one for us to marvel at! Until the early 1900s, grasslands covered one-fourth of California. Today, one percent of this habitat remains. Lost forever are 21,978,000 acres. We leave future generations a small fraction of the gift we were presented, but we have in our grasp the opportunity to add back a little that we lost!

ABOVE LEFT: ***LoKern.*** *The pot of gold at the end of this rainbow will be the yellow of the coming wildflowers. The rain this storm brings starts an incredible wave of green washing over the grasslands, a celebration of new life.*

ABOVE: ***LoKern.*** *Hoover's woolly-star.*

LoKern. *Weeks after the spring rains, the usually barren brown earth is a carpet of green. The plants take advantage of the mild temperatures and spring moisture by celebrating in color. This brief show passes quickly when summer comes.*

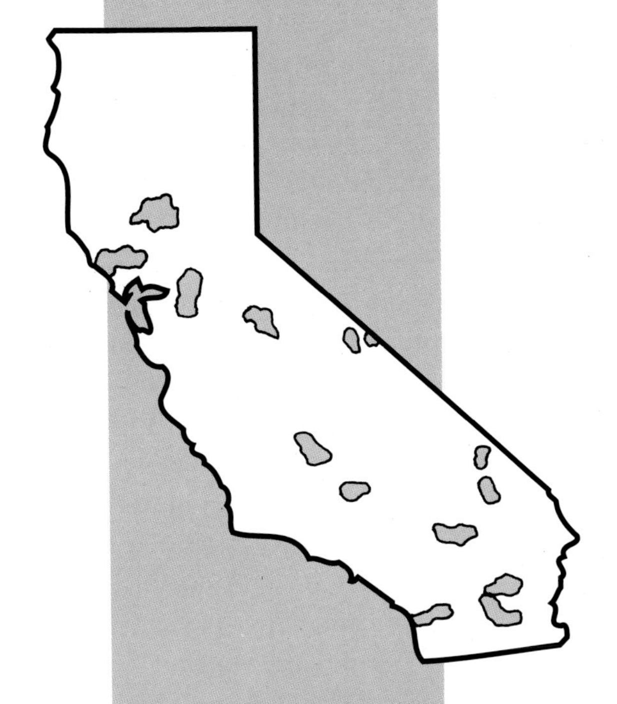

Fresh Water Marshes

The Interior Reflecting Pools

After riding hard for several hours, we came in sight of a considerable body of water, the shores of which were low, and covered for miles with the Tule rushes, —making a sight which was beautiful to look at.

James Capen Adams (Grizzly Adams), 1852

Some early explorers who came upon the freshwater marshes of California found them so large and impassable they cursed them. Others, such as Grizzly Adams, allowed themselves to get lost in the beauty and mystery of the marshes. More than four million acres of freshwater marsh confronted California's first settlers, and more than fifteen million waterfowl congregated in their waters, lifting off in flights said to darken the skies for hours.

The diversity of California's geology and topography allows for the formation of freshwater marshes in every corner of the state—in the mountains, the grasslands, next to the coast, and even in the deserts. They teem with life, from the smallest insect to the largest mammal, all depending on the flow of water for survival. Defining or labeling freshwater marshes is not simple. What they have in common is an aquatic ecology dependent on a freshwater source.

That does not mean that the water in these marshes is "fresh." The alkaline marshes of the Central Valley, Mono Lake, and the Salton Sea turn their freshwater input to water that is saltier than the sea. Basically, any marsh not fed by the ocean is considered a freshwater marsh.

Though they may be unnoticed by the untrained observer, the marshes have

LEFT: ***The Delta.*** *The delta is a maze of rivers, creeks, channels, and islands, supporting an incredible wealth of wildlife. At first, the birds and mammals are hidden among the rushes and shrubs, but if a visitor sits quietly for awhile, the wildlife will come out.*

seasonal changes. Those who frequently venture to their banks know the seasons by the birds inhabiting the waters. The male ruddy duck's bright blue bill rapidly pulsating on the water's surface signals spring. So does the red-winged blackbird, who flashes his red badge of courtship. Summer is signaled by the ducklings and goslings following their parents about the marsh. Fall is marked by the departure of the spring's young and the arrival of northern migrants coming to spend the winter.

Some marshes exist only in the spring or fall, filling with the winter rains, disappearing in summer. They may provide a safe place for waterfowl to roost and feed in the winter, then become a pond of multi-colored wildflowers in the spring before drying up in the summer. Their seasonal nature does not diminish their importance to the environment.

Understanding the vast variety of California's freshwater marshes and where they are found would constitute a book in itself. An idea of their grandeur and variety can be obtained by looking at some of the principal marshes making up the California landscape today.

Cache Creek. *The tule elk of the delta were driven to extinction long ago. This bull at Cache Creek is typical of what Grizzly Adams saw on the islands of the delta and in the tules surrounding the waterways.*

Gray Lodge. *Looking into the waters and islands of the refuge, the scene quickly transports the imagination back 150 years to a time when all the delta was intact. Among the waterways were islands where geese and ducks, cranes and shorebirds found sanctuary; in the rushes tule elk gave birth to their young.*

***Gray Lodge.** Once part of the huge delta network, the freshwater marshes of Gray Lodge now stand alone but still attract hundreds of thousands of waterfowl, geese, and cranes.*

SACRAMENTO DELTA

In this, we crossed an arm of the lake, and landed on a small wooded island, which was a place of birds indeed. There were birds in almost incredible numbers,—ducks, geese, swans, cranes, curlews, snipes, and various other kinds, in all stages of growth, and eggs by thousands among the grass and tules. There were also beaver's work in every direction; and we saw also elks in numbers, which fled into the tules at our approach.

James Capen Adams (Grizzly Adams), 1853

The delta region today is but a fraction of what it once was, but the marvels it holds still boggle the imagination. The delta is inland from the northern head of San Francisco Bay, San Pablo Bay to be exact. It is here at the delta that the mighty Sacramento and San Joaquin rivers meet along with the waters of the Mokelumne and Consumnes rivers.

The rivers deliver water from half the state's watershed, draining the entire Central Valley's runoff into San Francisco Bay. Along with the water, these rivers carry sand, mud, and silt, depositing them where they merge. Over a millennium, this has created the delta and has kept it maintained until recent times. Totalling seven hundred thousand acres, these islands were once enormous riparian forests or islands of vegetation. They formed marshes or entrapped water into back areas and created sloughs.

The vast richness of fish life in the waters of the delta goes unnoticed. Many species of fish depend on the delta as nurseries for their young. Others travel their waters to spawning grounds further up the river systems. Of these, the salmon is the best known.

Utilizing the Sacramento River, salmon travel from the sea through the delta to their final leg of migration. There are many upstream migrations of salmon, with the species being named for the time of their upstream migration. The fall-run, the late fall-run, the spring-run, and the endangered winter-run chinook salmon (numbering possibly as few as four hundred individuals) make up the major runs. The winter-run passes through the delta and upstream from November through June.

How can a fish so large or so important as a food crop become endangered? Water, a major component of the salmon's habitat, is being diverted, dammed, and/or polluted. As with many endangered species, the role of the salmon is very controversial. Water from the delta supports nearly two-thirds of the state's water needs, much of which is diverted to southern destinations. Some estimates say that as many as seventy-five percent of the salmon fry traveling to the ocean are killed by being sucked into the massive pumps moving this water throughout the state.

Delta Smelt

The four-inch, steel blue Delta smelt is endemic to the waters of the Sacramento and San Joaquin rivers and found nowhere else on earth. Once common in the upper estuary, it has lost its vital freshwater habitat. Inhabiting the now-murky waters of the delta, the smelt is in danger of extinction. Though still a scientifically contested fact, water diversion from the delta appears to be partially to blame for the smelt's decline. The large pumps that move the water out of the delta kill thousands of each year's fry.

The smelt replace their entire population each year; that is, all the adults die and the year's young take their place. They spawn upstream in fresh water between February and June. The eggs hatch after two weeks, and the fry are carried downstream to Suisun Bay where they grow in the null zone (where the salt water of the bay and fresh water from the river mix). Here they live until it is time to go upstream to spawn.

If that null zone should move out of the bay because of a flood of fresh water, such as occurred in 1982, there is a massive die-off. If the null zone moves upstream, which is the current problem, the smelt are more likely to be sucked into the pumps. It is estimated that one to two hundred thousand delta smelt remain; it is hard to believe they were so numerous in the 1850s they were a food source for the citizens of San Francisco.

The Delta. *Seemingly always on the wing, the greater sandhill crane and the lesser sandhill crane move about a considerable amount during the day. Their magnificent flights can be seen at almost any time around the delta.*

WOODBRIDGE ECOLOGICAL RESERVE

Getting lost in visions of past grandeur of the delta's bird life is easy to do at Woodbridge. Just south of Sacramento and literally next to Interstate 5, the marsh is constantly passed by motorists. Of the many species that come to Woodbridge to take refuge in its waters, none is as grand as the greater sandhill crane.

This threatened species nests in fresh-water marshes in northern California and southern Oregon, and winters in the Central Valley. The marshes are extremely shallow, providing just enough water for the cranes to make platform nests that look like small islands. The birds historically wintered throughout most of the Central Valley, but with less than three percent of their native habitat left, winter homes are difficult to find.

Woodbridge is on the eastern edge of the delta, its tule marsh replaced long ago by agriculture. The pastures of Woodbridge were lost as habitat before the Department of Fish and Game bought the area. Now water is purchased to flood the field and create a temporary marsh. Many believe the cranes depend on the marsh for food, but they actually prefer the grain and insects in the nearby fallow fields. They might nibble in the marsh, but it is the protection of its shallow waters that attracts the world's last five thousand sandhill cranes.

This haven is utilized by many other birds, not all of which depend directly on the water. Ducks, geese, swans, and shorebirds all frequent the quiet waters of Woodbridge. The threatened bank swallow feeds on the insects attracted by the water. Raptors like the black-shouldered kite, once listed as endangered, hover over the land to pick off unsuspecting mice scurrying about the marsh. The microcosm of Woodbridge is a reminder of past splendor, but it is also a present reflection of the Central Valley's rich marsh diversity.

ABOVE LEFT: ***Woodbridge Ecological Reserve.*** *Surrounded by agriculture, near the busy freeway, the calm, safe waters of Woodbridge provide an important and vanishing resource. This reclaimed freshwater marsh is vital to the survival of the greater sandhill crane.*

Woodbridge Ecological Reserve. *Standing in the warmth, this group of sandhill cranes chose not to take off to the fields. This is a mixed group; the tall cranes are the greater sandhill cranes and the shorter are lesser sandhill cranes.*

Lower Klamath. *A dark streak comes down from above and the ducks and geese on the frozen fields explode into the sky out of fright. This is how the bald eagles find the sick or injured birds, frozen in the ice, that they depend on for food.*

Lower Klamath / Tule Lake

Here, at any time of the year, you can close your eyes and open your mind and the sounds of life will come rushing in. From every direction, the lakes, marshes, fields, and skies above the refuges at Lower Klamath and Tule Lake are full of life. The richness of wildlife in the region captivates visitors, as the constant flow of life redefines the meaning of wild.

Klamath Basin extends into Oregon and supports six refuges. In 1908 the Lower Klamath Wildlife Refuge, on the border of California and Oregon, was the first refuge created for waterfowl in the United States. By then, much of the basin's 185,000 acres had been converted to agriculture. The remaining twenty-five percent—marshes, shallow lakes, and upland grasslands—supports over a million waterfowl in the fall, the largest wintering congregation of bald eagles in the lower forty-eight states, and is the nursery to countless thousands of waterfowl in the spring.

As the sun rises above the horizon, life in the refuge stirs. Hundreds of bald eagles are in the air, not in a flock, but flying over the area one by one on the prowl. Some weakened waterfowl died in the cold night; others chose the wrong stretch of open water as a night roost, and during the night it captured them in a frozen grasp. The eagles are out to find the birds who did not survive.

The bird life of Tule Lake and Lower Klamath is rich with diversity. Many raptors come from northern regions to spend the winter here. Most commonly seen are northern harriers and rough-legged hawks. The harriers spend vast amounts of time slowly gliding low over the marsh, aiming to surprise an unsuspecting mouse or weak duck and make a quick meal of it. Rough-legged hawks hunt from every perch above the surrounding fields.

Above Left: ***Lower Klamath.*** *Masses upon masses of greater white-fronted geese, Canada geese, mallards, wigeons, and pintails stage in the open fields.*

Left: ***Lower Klamath.*** *With the waters frozen, and the shrubs covered in ice created by the freezing temperatures and tule fog, it is hard to imagine this being an active refuge for wildlife. But this is a vital stopover for hundreds of thousands of waterfowl which are the main prey for the wintering bald eagles.*

Above: ***Lower Klamath.*** *Enduring the snow that lightly falls on its back, this great horned owl stays hunkered down in the grasses. Normally seen at dusk, when the great horned owls seem to come from everywhere to hunt at the refuge, the bad weather must have forced this one to take cover in the grasses where it is almost completely hidden from sight.*

The waterfowl put on a tremendous show, especially clustered together in an open grain field. Greater white-fronted geese, Canada geese, American wigeons, mallards, and northern pintails congregate tightly, watching for the prowling bald eagle or coyote. Honking, chattering nervously, and whistling, the group feeds or preens. Then, with only the slightest provocation, they burst into the sky. The crash of all their wings beating feverishly to lift off makes a tremendous noise. If you look for the cause of the commotion you can usually find a bald eagle diving low on the group, trying to find a weak or dead bird.

The most elegant birds to wing over the area are the tundra swans. These massive birds come, as their name suggests, from the tundra. They winter in a seemingly inhospitable climate which must seem like home to them. They announce their arrival by sound, which reaches the ear long before they are ever seen. Flying with their long, white necks outstretched, their grace in the air is matched by their presence on the water. Floating alone or in a flock, they constantly honk back and forth, reassuring themselves that all is well in their surroundings.

For the tundra swan, bald eagle, and all the others, the Lower Klamath and Tule Lake are safe. This unique slice of northern Great Basin freshwater marsh is protected from any type of development. Water for the marsh is plentiful and secure. The threat that these species might face would be on their breeding grounds, those they leave each fall to journey to this region.

Lower Klamath. *The rising sun gives way to a crack in the ice. This sliver of open water will soon swell with the arrival of waterfowl dependent on the water for safety.*

Tule Lake. *The great expanse of the tule basin, its large grasslands and freshwater marsh, has made it a major stopover on the Pacific flyway for centuries.*

Tule Lake. *Not an endangered species, the red-tailed hawk is an important indicator to the health of an ecosystem. At Tule Lake, the hawks fill the skies in the winter, foraging on the many meadow mice and voles. This individual has just come down on an unsuspecting vole.*

Tule Lake. *The frozen grasslands take on a life of their own. The tule fog freezes on the shrubs each night and transforms the barren landscape into an ice palace for the imagination.*

Tule Lake. *In the snow-dusted fields of the refuge, the Canada geese are able to find food and safety. At this moment they are up on their toes looking about, probably because of a prowling coyote over the rise.*

TECOPA MARSH

***Tule Lake.** Inhabiting Sheepy Ridge, named for a species now extinct in the region, this mule deer takes advantage of the available shrubs left uncovered by winter snows.*

Finding Tecopa March on the California map takes a very keen eye. Tucked away on California's eastern border, southeast of Death Valley, Tecopa's forty-five hundred acres go unnoticed by most. This extremely fragile, desert, alkali marsh was inhabited by only a few settlers, starting with the first Spanish explorers, then borax miners in the late 1800s.

A large flood plain for the Amargosa River, Tecopa Marsh is wet all year round. There is always some surface water collecting in the lowest lying areas, but the marshy pools come to life in the winter and spring when rain refills the alkaline salt beds. The surface of this massive mirage-like lake is covered with thousands of waterfowl that fly far out in the desert to stop over during their annual southern migration. Gazing at the tan-colored mud in the summer, imagining such bird life on the marsh is difficult at best. Between March and October, a visitor to the marsh sees only the hardiest of marsh dwellers. The tall great blue heron, showy American avocet, noisy killdeer, nervous mallard, and shy snipe, along with a sprinkle of sandpipers here and there, make up the bird life on the marsh during the crushing heat of summer.

Those precious waters that the Amargosa River delivers to Tecopa originate in Nevada near Ash Meadows. The Amargosa is not much of a river; most of the time its course is spent underground, only emerging when the geology forces it to the surface. The locales where the Amargosa River does surface are like any legendary life-giving oasis in the desert. Tecopa Marsh is no exception.

The California desert is noted for being home to a specialized group of fish, the pupfish. These prehistoric leftovers lived in the large lake that once covered much of California's desert. As this lake dried up over the years, the pupfish were isolated into small lakes, ponds, and pools. The Tecopa pupfish evolved to live in the Tecopa Marsh area. Because of degradation of its habitat, both the loss of water and its pollution, the Tecopa pupfish is now extinct.

The Amargosa pupfish and Amargosa Canyon dace depend on these same waters. These two fish are not yet listed as endangered. As more development in the area and upstream damages the water flow, these two fish will go the way of the Tecopa pupfish.

Walking over the land at Tecopa, the white of the alkali soil clashes with the reds and browns of the volcanic earth and causes one to wonder what plants could ever grow here. In areas of this soil brine where moisture from springs or the Amargosa River loosens the soil, a small plant emerges.

Tecopa Marsh. *In the open expanse of the Colorado Desert amongst the creosote scrub, the Amargosa River winds its way down to the Tecopa Marsh. Along its path, the river brings water to this fragile habitat while providing a critical element to the desert's inhabitants.*

The Amargosa niterwort, no larger than your thumbnail, is an endangered species. Found only at the Carson Slough north of Tecopa and at Tecopa, the lowering water tables have pushed it to the edge. It can be likened to many of the desert's endangered inhabitants. Crusty, small, unique, and difficult to find, many wonder why this plant should be protected. Any living thing that can eke out an existence in a landscape such as this deserves the right to keep on doing so. Who are we to decide when that existence should come to an end?

Tecopa Marsh is not protected, not its inhabitants nor the precious water that brings them all life. Plans have been proposed to tap the Amargosa River's water source near Ash Meadows, possibly drying up much of the wetlands along the river, including Tecopa Marsh. The town of Tecopa on the edge of the marsh is the proposed site of a new large resort and golf course. Predictions of this freshwater marsh drying up and blowing away are all too real!

Carson Slough. *North of Tecopa Marsh on the Amargosa River, Carson Slough is an anomaly in the barren desert. It is also the home of a tiny endangered plant, the Amargosa niterwort.*

Amargosa Vole

The Amargosa vole is a small mammal, possibly one of California's most endangered. It evolved to live in only one place, the freshwater marsh of Tecopa. Creating small runways in the bullrushes and salt grass along running water, this little creature ekes out an existence where the summertime temperature exceeds 100 degrees.

Once thought extinct, it was rediscovered in the mid-1930s. Biologists have been unable to learn much about the Amargosa vole's reclusive population. Between 1975 and 1977, at least twenty individuals were known to exist. In 1988, less than ten were found still inhabiting a small portion of the existing marsh. In 1992, we knew for a fact that one Amargosa vole still lived in the Tecopa marsh.

The vole pictured is the last one seen by biologists. Known locally as "swamp rats," they were once caught by local boys. Many will ask why this species is so important. Like the fragile habitat where it lives, the Amargosa vole is part of the puzzle. Without every piece, the picture will never be complete.

***Fish Slough.** In the background are the White Mountains, as ancient as Fish Slough and just as vital. In fall, when the rushes and other vegetation turn brown, the life of the marsh slows down in preparation for winter.*

FISH SLOUGH

As unusual as any oasis in the desert is, Fish Slough is even more so! Just north of Bishop, its thirty-six thousand acres were designated as an Area of Critical Environmental Concern (ACEC) in 1982. This means that Fish Slough is recognized as an area that needs to be maintained and enhanced, and its unique resource values protected. It is unique because it is is a year-round, freshwater marsh in the high desert.

A natural spring on the Owens Valley floor feeds and replenishes Fish Slough's marshes. The volcanic tableland geographic forces have given life to Fish Slough. The wetland habitat exists because water passes through and is caught in the tableland's gigantic underground water reservoir. The water comes to the surface at Fish Slough in the form of springs, which can be seen bubbling up in many of the marsh ponds. The climate around the slough is arid. It receives less than six inches of rainfall annually with summer temperatures past the century mark, but the tableland formation is able to keep the slough wet all year.

These same volcanic formations give Fish Slough its unique look. Most marshes

Fish Slough. *The slough is an important breeding area and this pied-billed grebe is just one of many species taking advantage of the water's safety. The young grebe has a very striking striped pattern at birth, which slowly fades as it grows older.*

Fish Slough. *Climbing the high ridge that is the east wall of the slough, this aerial view discloses the many secret pools making up the slough. The crystal blue water, some of it warmed by hot springs, is home to many waterfowl, wading birds, and fish.*

are large, flat areas, perhaps with mountains in the distance. Fish Slough has a large eastern wall (the remnant of a volcano) and more ledges to the west, creating a narrow valley that houses the slough.

The combination of water and desert environment creates unique species adapted to these conditions. In plants, such specialization can be seen most markedly. The Fish Slough milk vetch, being considered for listing as threatened, holds on to a precarious niche. This species was only discovered in 1974 by a local botanist; it lives nowhere else in the world. The alkali Mariposa lily and Mono buckwheat are other plants in the slough that are threatened with extinction. They each depend on a certain location where the combination of water, soil, and sun exposure is optimal for their survival.

Of all its inhabitants, none depend on Fish Slough for their very existence more than the Owens pupfish. Though it now resides in a number of ponds, Fish Slough saved this species from extinction.

Owens Pupfish

In 1859, an U.S. Army captain discovered this queer little fish in a number of wetland areas of the Owens Valley. There are notations of large schools of pupfish in the waters in and around Bishop, but the water diversions in the 1920s and 1930s shrank or wiped out many of those ponds. The pupfish was thought to be extinct by 1948.

Years later a couple were found to exist in Fish Slough and in 1964, a small group of two hundred fish had to be moved because their habitat was being dried up. Men and buckets came to the rescue, moving the pupfish to other ponds in the slough. Now the Owens pupfish is thought to be in stable condition and has been introduced to other ponds, although Fish Slough remains one of its largest strongholds.

Pupfish typically survive in warm, alkali water. Low oxygen levels in the water are also common. They feed on small insect larvae and decomposing aquatic plants. No more than two inches long, the male pupfish aggressively protects his territory from other males. The male can have a blue tint (missing on the female) to his back and sides.

Sharing these waters with the pupfish is the Owens tui chub. A funny little fish with a funny name, it is endangered for the same reasons as the pupfish. One threat to both, which is not related to the drying of their habitat, comes from other fish. Bass and trout introduced to this habitat eat the two endangered fish and their usual food. Efforts are under way to exclude these large predators from the waters of the pupfish and chub but are slow going. Some of these predatory fish are thought to have been introduced by irresponsible fishermen.

Fish Slough is unique in another manner. It is managed by not one, but at least five different entities: The U.S. Fish and Wildlife Service, University of California, Bureau of Land Management, California Department of Fish and Game, and City of Los Angeles Department of Water and Power. The latter two have joined to establish the Fish Slough Native Fish Sanctuary. Because of this, the future for Fish Slough is more secure. But like any water in the desert, diversion from miles away can affect the health of the freshwater marsh of Fish Slough.

ABOVE: ***Fish Slough.*** *Hundreds of miles from the ocean, it always seems odd to see shorebirds in the desert. But like the greater yellowlegs, many shorebirds depend on this desert freshwater marsh in the winter.*

RIGHT: ***Fish Slough.*** *Hiding in the rushes, coming out to greet the morning sun, the least bittern is one of California's shiest birds. These birds spend much of their life in the rushes, traveling through them by grasping one after another. When they feel threatened, they point their beaks to the sky, then slowly sway to imitate the breeze moving the rushes.*

Salton Sea. *One of the major areas at the sea for birds is Wister, a Department of Fish and Game wildlands area. Hundreds of species winter and reside here; in this pond, great egret, snowy egret, white-faced ibis, and forester's and gull-billed terns are in a feeding frenzy.*

Salton Sea

Thought of as a dead sea, the Salton Sea is an anomaly in the freshwater marsh habitat schematic. A huge body of water filled by freshwater streams and rivers, it takes its freshwater inflow and turns it into water three times saltier than the ocean. West of Palm Springs, this twenty-two-mile-long body of water in the Colorado Desert is below sea level. But what is spectacular about this sea is the life it supports.

Without an outlet except for evaporation, the Salton Sea has a water problem caused by its rising salt content. Its water level has been increasing for a long time as runoff from agriculture spills into its salty brine. With that runoff come more salts, leached from the soil along with selenium and other trace metals. Also, polluted water from Mexico is deposited into the Salton Sea. So the sea's problem is getting fresh water that is unpolluted.

There are many fragmented freshwater marshes along its edges; the largest is Wister, a state wildlife area used extensively in the fall by waterfowl. During the spring, thousands of great and snowy egrets take advantage of the calm water. White-faced ibis and the endangered wood stork also prowl the water for a meal. Hidden in the bullrushes around the edge of the pond is a secretive bird, the Yuma clapper rail.

In the heat of summer, Salton Sea hosts such unique birds as gull-billed terns and black skimmers. White and brown pelicans, snow geese, black terns, yellow-footed gulls, cormorants and ducks by the thousands, herons and cranes all take advantage of this massive body of water. Some of the only heron rookeries existing in a desert are here.

Salton Sea does not have a secure spot on the California landscape. With a majority of Californians living in the desert region known as Southern California, water will always be a needed and contested commodity. This, combined with agricultural interests putting even more pressure on a limited water reserve, places Salton Sea on the edge of destruction. Although there is no such thing as an endangered species list for habitats, maybe there should be!

Yuma Clapper Rail

California's only freshwater clapper rail was once found all along the Colorado River, but now inhabits only a small track of the old Colorado River channel and Wister's marsh. At some point, this chicken-sized bird flew from the Colorado River, discovered the Salton Sea's marsh, and inadvertently secured its own survival by starting a new population. Its native habitat on the river was subsequently destroyed as the river was tamed and its waters diverted for agriculture. The one hundred or so birds that live at Wister are estimated to be one-fourth of the world population. The Yuma clapper rail is very much like its saltwater cousins biologically, which includes being on the edge of extinction.

Mono Lake

This is the most remarkable lake I have ever seen. It lies in a basin at the height of 6800 feet above the sea. Like the Dead Sea, it is without an outlet. The streams running into it all evaporate from the surface, so of course it is very salt—not common salt.... When still, it looks like oil, it is so thick, and it is not easily disturbed. Although nearly twenty miles long it is often so smooth that the opposite mountains are mirrored in it as in a glass.

William H. Brewer, 1863

Mono Lake, nestled up against the eastern Sierra Nevada in the Great Basin, supports the breeding population of one-quarter of the world's California gulls and half of the world's Wilson's phalaropes. Though the bird life diversity of Mono Lake is not on a grand scale, their sheer numbers are! What brings such high concentrations of birds to this desert lake? It is the water and the inhabitants of the salty brew.

The brine shrimp and brine fly in Mono Lake are tiny, but what they lack in size they make up for in numbers. The flies can cluster so thickly on the shore you can't even see your shoes. The shrimp are so plentiful that they are harvested and sold for fish food.

Mono Lake's two extinct volcano islands serve as a safe refuge for nesting. Before water diversion took away the fresh water flowing into Mono Lake, clouds of gulls could be seen rising over the islands in the spring and summer. When the water level of the lake was lowered, land bridges were exposed, connecting the islands and their nesting colonies to mainland predators. The safety of the islands—when they are islands—in concert with the abundant food source, is what attracts the hundred of thousands of birds to Mono Lake.

For one of the largest freshwater marshes in California, Mono Lake's biological story is one of the smallest. The chain of life necessary to sustain such an incredible natural wealth can easily be seen. Yet Mono Lake's problems have been national news for over a decade. Those streams that were diverted are, by court order, flowing again, but the drought has had its effect. Like many endangered species, when things are on the edge, stress such as a drought can be the final push. Mono Lake's future is brighter with the rains of 1992, but until the lake rises once again to a level that protects its wild heritage, concern for this natural wonder is warranted.

Mono Lake. *Half of the world's population of Wilson's phalaropes come to Mono Lake each year. The lake's brine flies and shrimp attract thousands of birds to its waters.*

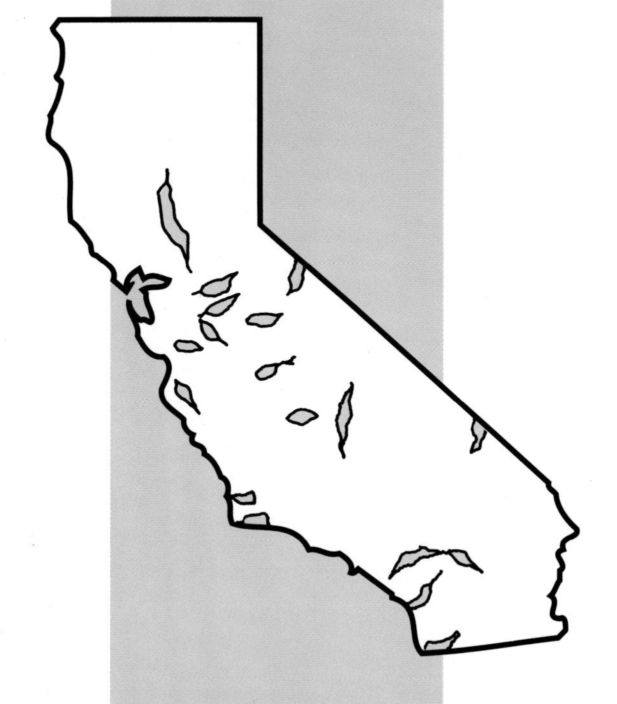

RIPARIAN FORESTS

The Ribbon Forests

The great trees themselves . . . are wonderful and stately enough, the tall, tapering shafts rising in superb grace and power, flecked with purple and gold along their fluted channellings. A forest of their kind surrounds them, mingled with a few other species, and the clear, bright river ripples or steals along as seductively as river can do . . .

J. Smeaton Chase, 1913

The riparian forests of myth are home to goblins and ghouls, living in the dark recesses of the forest's inner realm. Life in these forests has long been a secret, except to those who ventured in and discovered their marvels. Their rich diversity staggers the mind; even the small examples we have left are a testament to that. But, as with most riparian forests in North America, much of California's were lost before they were even described.

LEFT: ***Mono Basin.*** *The green foliage of spring sets the cottonwoods of the riparian track apart from the dark green of the surrounding chaparral. This narrow corridor is the nesting ground for sixty species of birds, which is typical for a riparian forest.*

California has lost eighty-nine percent of its riparian forest tracks in the last 150 years. Riparian forests historically covered one million acres in the Sacramento and San Joaquin valleys alone. Found throughout the state from grasslands to deserts, these forests depend on water, from permanent as well as seasonal sources. Diverting, damming, taming, and channelizing the state's water courses, and converting land for agriculture and development, are the main causes for the loss of our riparian forests.

The riparian forest cannot be generally classified by just one tree species. Throughout the state these forests are dominated by California sycamore, valley oak, bigleaf maple, California black walnut, Fremont and black cottonwood, or a variety of willows. Many forests are a combination of these trees; they support a rich and varied shrub growth that in turn supports the many species of wildlife that live in the riparian forest.

The trees of the riparian forest are determined by the locale and water supply. The thick forest of the Sacramento delta has oak, the Central Valley a combination of oak and cottonwood, while the south coast has willow and cottonwood. With each change above comes a subtle difference below in shrubs, plants, and wildlife. The connecting thread in this forest is water, dictating its very structure and existence. Paralleling the water course, mirroring its every turn, the riparian forest's appearance from above is that of a ribbon of green.

Mono Basin

It is a fine grassy plain, with here and there a gentle green knoll, with a few dry creeks or alkaline ponds, and one fine stream, the Santa Clara River running though it. We stopped for an hour on its bank and rested our mules, lunched and refreshed ourselves in a grove of cottonwoods which came nearer to a forest than anything I have yet seen here.

William H. Brewer, 1860

Tucked away in the corner of the Santa Ynez Valley, no more than fifteen miles away from Santa Barbara as the vireo flies, Mono Basin stands out as a green gem in a sea of brown chaparral. Truly defining a ribbon forest, the cottonwood-willow riparian forest of Mono Basin tightly follows the course of Mono Creek to Gibraltor Reservoir. Along its entire course, the green belt provides nesting habitat for the rich diversity of birds that migrate here to raise a new generation.

The winter rains start the process, bringing water and new growth to the forest's many plants. In a really wet winter, the forest receives its needed cleansing from the creek's heavy rush of water, scouring away much of the old debris that gathers over the years. As a result, a rich profusion of shrubs sprout, covering the forest floor with green growth. In March, the cool temperatures dissipate into the heat of summer.

Looking down on the forest, only the thick canopy of willows and cottonwoods greets the eye. This roof of green leaves shades the many shrubs of the forest from the temperatures of ninety to one hundred degrees. These shrubs, such as the mugwort, are vital to the many birds which nest on or near the ground.

Standing in the forest itself, you see the rich shrub layer rising three feet above the forest floor. From the shrub tops to the bottom of the tree canopy is an open flyway where the birds forage. In the spring, the first to return to Mono Basin are the yellow-breasted chats. They announce their arrival with boisterous song and a comical aerial dance. Soon, yellow warblers, warbling vireos, willow and ash-throated flycatchers, and white-throated swifts join the chat. A variety of other species pass through the basin, stopping over long enough to gain energy to continue their migration.

The arrival of Mono Basin's most notable resident, the least Bell's vireo, signals the official start of nesting season. More than sixty-six different bird species compete for space and food along the mile-long corridor.

Least Bell's Vireo

This drab, little bird once inhabited much of the riparian forests of south and central California, extending its range into Yosemite Valley. With the vast destruction of the forest, the vireo slowly disappeared until now it safely resides in just three riparian forests in southern California. Strictly a spring and summer visitor, here to raise its young, the vireo has adapted a unique strategy to survive and continue its race.

The males arrive first, setting up territories that they defend with their crazy, non-lyrical song. These territories are normally the exact same ones the males had the year before, down to the last shrub and branch. While singing to defend his territory, the male also tries to attract a mate, possibly the same one as the year before. If successful, the nest he builds is soon filled with four eggs that the female incubates. Within sixteen days the eggs hatch and fourteen days later they fledge. For most birds, this would be the end of their nesting duties, unless the first nest was unsuccessful. But the vireos seem to double clutch quite often, repeating the whole process, and doubling the number of young they add to their species each year.

The vireo was the first bird residing in a riparian forest to be listed as endangered. The listing was fiercely contested, and millions of dollars in construction projects halted or were stopped altogether. The storm over this small, gray bird was one few had seen until that time. Currently other species are in the limelight. But it seems criminal that much of our natural heritage goes unnoticed until it winds up in the headlines.

In April the cottonwood trees drop their seed pods; encased in cotton-like puffs, the pods float to the ground to germinate new trees. Many of the bird species, like the vireo, use this natural seed down to line their nests.

Surrounding the riparian track is a very important component of the breeding birds' success. In Mono Basin's case, it is the chaparral habitat. Even with the immense amount of insect life to support these birds, many have to venture into the chaparral to forage. This codependence on neighboring habitats is not unique to Mono Basin's riparian forest, but it is usually an overlooked aspect of the forest's ecology.

This textbook role model for a riparian forest was almost lost not long ago. The need for water on the southern California coastal plain has dried up or flooded out over ninety percent of its historic riparian forests. The listing of the least Bell's vireo, and the basin as its critical habitat, saved Mono Basin from destruction. The ironic piece to this puzzle is that the debris dam, built to keep silt out of the reservoir, created the forest decades ago. Construction of the reservoir destroyed the original forest and started the whole downhill process. For now, this population of least Bell's vireo and riparian habitat is safe.

Big Morongo

Water in the desert constitutes an oasis, and the riparian forest and marsh of Big Morongo is truly that. This canyon was once a large plateau where two rivers flowed. A down-faulting occurred along its fault with such friction, it melted the two-billion-year-old granite (some of the oldest rock in California), sealing it and making it impervious to water. The two rivers continued to flow, but became a large lake in this newly created canyon. The rivers carried debris, and over a long period of time buried their own paths, going underground and forming the Morongo Valley.

Nestled between the Mohave and Colorado deserts west of Joshua Tree National Monument, water still flows twelve months of the year, rising in the six-mile-long Big Morongo to nourish its riparian forest. Big Morongo is among the ten largest oases in California's desert and one of the five largest tracts of cottonwood-willow habitat in the California desert. It also has the greatest variety of flora and fauna of any of the twenty-two oases in the Mohave Desert. Finding refuge in its lush green foliage are over 270 species of birds and 355 different plant species.

Growing in the desert hills surrounding the riparian tract is the familiar creosote bush. It brings many of the typical desert species in contact with those of the forest, adding to the forest's diversity.

Species such as the desert bighorn sheep come down from the hills to take advantage of the water, returning to their high sanctuary once they have obtained their fill. Many desert birds do likewise, venturing into the oasis for a quick moment before returning to their dry world. The mingling of desert bird species such as verdin and phainopepla with marsh wrens and willow flycatchers is common. Birds are the main attraction of any riparian forest, and Big Morongo has one of the finest collections. Twenty-nine species of wood warblers visit the forest. All the water bubbling up to support the willows and cottonwoods also supports a freshwater marsh. In the forest, this refuge attracts some of the rarest of marsh dwellers. Sora, Virginia, and black rails have been seen cruising about the marsh plants.

The water that brings life to Big Morongo is by no means secure. As more and more development occurs upstream from Big Morongo, the water table is lowering, affecting the water flow to the riparian forest. It is already reduced in size from historic accounts but still supports an incredible gathering of life. Foresight in planning is the only guarantee that this desert oasis will always remain a green ribbon in a desert of sand.

Big Morongo. *The Mohave Desert's sage grows right to the edge of the riparian forest, which stands out as a green ribbon in the background. This diversity of habitats is what makes this forest so rich with wildlife.*

Consumnes River. *This meeting of river and oaks is a true oak riparian forest. Scenes such as this can only be found in a few locales in the Central Valley, since most of this forest has been converted to other uses.*

Consumnes River. *The steel glare of a Swainson's hawk is hard to shake off, but this majestic flier of the Central Valley is in danger. It nests in the tops of trees in the riparian forest, and the loss of that habitat has put the Swainson's on the edge.*

CONSUMNES RIVER

A person who has not experienced the influence of an early, calm, summer morning, with the heavens lighting up in a crimson glow above, and the birds wakening into song around him, may perhaps imagine, but cannot feel the beauty and joyousness of the scene.

James Capen Adams (Grizzly Adams), 1852

These words paint the scene of a dawn that could have been found anywhere in the San Joaquin Valley in the 1850s. For Grizzly Adams, the oak riparian forest was the home of many of the species of birds and mammals he sought. Conversion of the valley floor to agriculture and the taming of rivers to supply water and protect real estate has meant the slow extinction of California's "rain forest."

The Consumnes River near the town of Woodbridge still harbors forest that Grizzly Adams might have once prowled. Unlike the stereotypical riparian forest of dense thickets and mysterious sounds, the Consumnes forest is grand and stately, made up of regal giants standing guard over its river banks.

The Consumnes is the only major river left on the valley floor that is not dammed. Hours after high tide reaches San Francisco Bay, the waters of the Consumnes rise too, slowly and almost imperceptibly, as they always have.

The Consumnes is a major river, and when the snows melt or the rains pour, it can flood its banks, creating seasonal freshwater marshes outside the riparian track. This doesn't destroy habitat, but only enhances it and makes it valuable to more species.

The valley oaks which make up the riparian forest of the Consumnes are believed to be the highest quality and largest remaining stand in the state. The oaks began to vanish after the gold rush when their habitat was recognized by farmers as being the most fertile and best watered. The trees were used to build and heat the new farms, and were sent to other regions for firewood. During the process, ninety percent of the oak riparian forest of the valley floor was lost.

These majestic giants guard a quiet habitat, with a slow and easy pace to match the river. In the marsh next to the oaks might be fifty lesser sandhill cranes roosting in the noontime sun, and nearby a pair of river otters play in the reeds. Frogs and turtles sun on the banks of the river, splashing in when disturbed. There is none of the hustle and bustle of a typical riparian forest.

Overhead a Swainson's hawk screams at an intruding red-tailed hawk. The dark, handsome Swainson's hawk is a threatened

Consumnes River. *The American wigeon uses the river and the freshwater marshes. This majestic waterfowl is common in healthy habitats and takes advantage of the seasonal wetlands and their rich forage.*

species because it nests in the diminishing Central Valley riparian tracks. It builds its nests high in the oaks and feeds on small mammals that abound in the forest. One of our few migratory raptors, the hawks spend their springs here and their winters in South America.

The list of wildlife in the Consumnes may not be as lengthy as that of other riparian habitats, nor may the forest attain the great blaze of green of a cottonwood forest. But there is no doubt this is a vital and endangered habitat. Loss of the forest from the Central Valley has signaled the endangerment of a number of our species. There is no guarantee that this loss will not continue. Other than the Consumnes, remaining examples of this habitat are threatened with extinction right now. Whether it is from pollution, loss of water, or deforestation, these riparian tracks might not be here in the next century.

San Luis National Wildlife Refuge. *The tule elk once filled the waterways of the Consumnes, taking advantage of the rich, sweet grasses growing there. The elk are now gone from here, a missing link in this otherwise historic ecosystem.*

Great Falls Basin. *Just south of Death Valley, the basin's rocky pockets provide shelter and water for a number of riparian forests. It is hard to imagine that such a barren-looking landscape could harbor such a specialized riparian bird as the Inyo California towhee.*

Great Falls Basin

The most unlikely candidate for riparian forest recognition must be this patch in Great Falls Basin. High in the Argus Mountains outside of Trona, a massive collection of boulders and rocks forms a natural fortress. The only way up inside its walls is a steep burro trail, and after miles of hiking in the desert to reach the top and still seeing only piles of rock in the shape of a huge bowl, it is easy to think this highly specialized riparian forest must be a myth.

The makeup of Great Falls Basin's riparian forest is like no other in the state. In pockets of decomposed granite with little soil or debris, water emerges from underground springs, and dense, shrubby thickets of willow take hold. Arroyo and yellow or narrow-leaf willow are the major plants that characterize these riparian stands. Where temperatures can easily reach 100 degrees sixty days a year, the willows stay green and the water flows.

What makes these riparian stands so important? They are the only home for the endangered Inyo California towhee. One of twelve subspecies of California towhee (formerly known as brown towhee) common to most backyards, its numbers are fewer than 175 individuals. That's the total world population and they all reside in these small pockets of riparian forest. They require four to five thousand feet of riparian track for nesting and shelter.

Very unlike the towhee in your bird feeder, the Inyo California towhee is extremely shy and evasive. The dense riparian habitat provides privacy for the towhee to nest and forage. But the towhee requires riparian tracks near these desert hillsides so they can also forage in that vegetation. This bordering habitat dependence is common to many riparian bird species, but the extreme contrasts of this environment are unlike any others in the state.

How can a bird so shy, and habitat so specialized, existing so far away from any development, be in trouble? The answer hinges on the water that provides life to the forest. Gold mining and other activities lowered the water table and dried up many of the springs. Those that still flow do so at a decreased rate. The shallower water attracts feral burros, which wallow in the water, smashing down and destroying the fragile riparian structure. The fate of the Inyo California towhee is in doubt over the long term; it may disappear from this oasis in the barren wilderness unnoticed!

Great Falls Basin. *Not your generic California towhee, the Inyo California towhee is a very shy bird living in the heart of the Mohave Desert. Because it depends on the meager riparian forest of the basin, its future is in doubt.*

Kern River

High in the Sierra, where pines are thought of as being the only forest, is found twenty percent of California's remaining riparian forests. East of Lake Isabella, the south fork of the Kern River was spared total destruction, preserving five thousand acres of magnificent Sierra riparian forest.

The Kern's riparian forest is a complex plant community. It lies where the Sierra, Mohave Desert, and Central Valley plant communities meet. On the slopes overlooking the forest, desert plants like beavertail cactus can be found. Salt and wild rye grass, species typical of the Central Valley, grow in the meadows. The diversity sets this forest apart from all others.

In this botanically rich environment, much of California's rare and endangered wildlife finds a place to survive. One plant, the alkali mariposa lily, lights the riparian floor with its blossoms in May. Many flowering plants inhabit the forest, so many that the area is the site of an annual Xerces Society butterfly count. Rare butterflies, such as the San Emigdio blue and the eunus skipper (thought to be extinct until rediscovered here) depend on the forest's blooms. There are a total of eighty-one species of butterfly found living in the riparian forest.

The western pond turtle, currently under consideration for endangered status, lives in the Kern's waters. The turtle hibernates from November to February in the mud at the bottom of ponds, but much of its watery habitat has been lost or degraded.

The rich variety of birds that dwell here also distinguishes the Kern riparian forest. Two of its threatened bird species are the western yellow-billed cuckoo and the willow flycatcher. This is one of the last riparian forests in California that meets their requirements for nesting.

The future of the Kern riparian forest is one of the bright notes for our endangered habitats. The Kern's plight was noticed early, and a dedicated group headed by the Nature Conservancy made a commitment to preserve and enhance its natural wonders. Though it is only a very small slice of its past size, its majestic stature in our landscape reminds us of the good we can do and the success we can have when we are involved with preserving our natural heritage.

Kern River. *The green belt between the hillsides of brown is a unique riparian forest home to the willow flycatcher and western yellow-billed cuckoo. While not a land of many colors and designs, species such as the cuckoo and flycatcher depend on it for their survival.*

Los Banos Grande

Los Banos del Padre Arroyo de la Cuesta, the original name for Los Banos Creek, was given to the area by padres from Mission San Juan Bautista. In 1805, the padres ventured over the Gabilan Mountains towards the San Joaquin Valley; en route they stopped and bathed (banos meaning bath) in the creek's cool pools. The original name was shortened again and again by the Europeans settling the valley until the creek, canyon, and town all became simply Los Banos.

Explorers of our time now venture to this canyon to bathe not in Los Banos's pools, but in its rich natural wonders. Its riparian forest is not the typical wall of green shrouding all its inhabitants, but Los Banos Grande, last of our remaining oak-sycamore riparian forests, is itself very endangered!

Walking west up the canyon from Los Banos Reservoir, the rolling foothills covered in grasses hide the grandeur that lies beyond.

Los Banos Grande. *Amongst the natural formations of lichen-covered asphalt, following the course of an ancient stream, Los Banos Grande is one of the last sycamore riparian forests in California. It supports such endangered species as the San Joaquin kit fox as well as rare species like the prairie falcon.*

These grasses are a home to the San Joaquin kit fox. The wide open spaces, abundant rodent life, and lack of human pressure allow this population of fox to live as did their ancestors, who probably witnessed the coming of the padres.

Further up the canyon, the way becomes narrower as the sycamores and oaks appear. In the spring, each turn of the canyon reveals wildflowers of differing color, shape, and size. The majestic California poppies pale next to the wild onion, blue dicks, and fields of tancy-leaved phacelia.

The further you venture up the canyon, the more the region's geology tells its past. Fault lines, geologic uplift, and volcanic activity were all major forces in creating Los Banos Grande. Bizarre shapes of asphalt-based strata outcroppings characterize the canyon. Lichens and mosses cling to them, adding color to their unique form.

With the gift of the winter rains, Los Banos Creek fills, bringing to life the frogs inhabiting its banks. One is the red-legged frog, a candidate species for listing. Much of its riparian habitat has been lost, its shaded waters nearly gone and with them, the frog. The western pond turtle shares this habitat, too.

Evidence of the rich diversity of life in this forest can be found in the skies. Overhead, golden eagles use the afternoon

Western Yellow-Billed Cuckoo

Known as the rain crow because of its habit of singing just before it rains, the western yellow-billed cuckoo has little to sing about these days. A 1987 estimate stated that thirty-one to forty-one pairs were all that were left of the historical thousands that flew across California.

The loss of its habitat, which decreased from one million acres to less than one hundred thousand acres, has reduced this magnificent, secretive bird's nesting sites to just a few riparian forests. The Kern, Sacramento, Feather, Amargosa, Santa Ana, and lower Colorado rivers are all it has left. The cuckoo is under careful study to prevent it from disappearing from our landscape. Part of its problem lies in pesticides that have reduced its egg shell's thickness by twenty-one percent. Another problem, associated with its loss of habitat, is that the cuckoo does not have stopover locations during migration. The loss of the cuckoo in Oregon and Washington has been attributed to this.

The Nature Conservancy, at its Kern River Preserve, is helping the cuckoo by restoring riparian forest to areas where it was destroyed. This active program is not an overnight solution, but is a positive step towards the future.

thermals to scan the forest for a meal to feed their young in a nest high on the ridge. Two prairie falcons take advantage of the same diverse forest, building their nest in a small hole in the cliff above the forest. The eagles prey on the small mammals in and around the forest, while the prairie falcons hunt the forest birds. Both of these predators depend on the rich life in the forest to raise their own young, year after year.

The remaining grandeur of Los Banos Grande's riparian forest might go the way much of it has already gone, under water. The riparian forest that once extended all the way to the valley floor was lost long ago. Plans are now being discussed to flood the remainder of this magnificent and disappearing habitat. This is a canyon which has already experienced extinction, not just as a riparian forest, but as home to a Native American civilization. The first to come enjoy and depend on the riches of Los Banos Grande were led off in ropes to another land; their town and burial sites are now underwater. This canyon, Los Banos Grande, has witnessed the passing of time and has earned the right to continue doing so, long after we have bathed in its serene beauty.

Los Banos Grande. *The coastal tidy tips sprinkle their yellow greetings on the spring field of green. The riparian forest, no matter if it is cottonwood, oak or sycamore, is at its finest in the spring when every shade of green is displayed.*

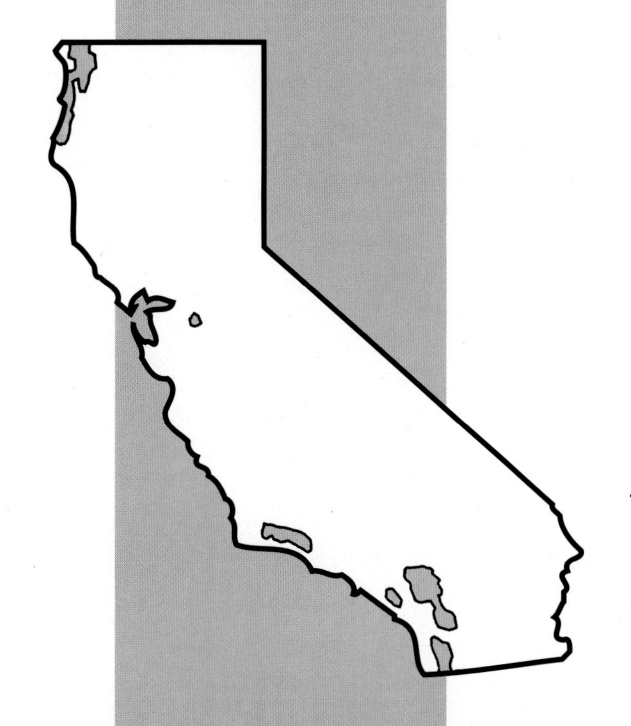

OLD GROWTH FORESTS

Sentinels of Time

The Redwood tree is the main-stay of California. The supply is inexhaustible, but nature has been sufficiently capricious to make them most abundant in very inaccessible spots, while the level plains are covered with a short-grained dwarf-oak, serviceable only for firewood. But however steep the mountains, the California redwood has to fall and to be forested to the use of man, and when a steam saw-mill gets perched upon a mountain-top, the romance of the forest is gone; its silent grandeur no longer awes the mind; and the trees, whose size and beauty caused such deep impressions and such grave reflections, fall into insignificance . . .

Frank Marryat, 1855

California's old growth forest, its inspiring pines, sycamores, oaks, and redwoods, stood over the state long before recorded time. The noblest of oaks, the tallest of redwoods, the grandest of sycamores, and the thickest of pine forests covered the landscape from the ocean to timberline in the mighty Sierra Nevada. They encompassed every color, size, and shape the imagination can picture and provided homes for thousands of other forms of life.

The lure of California's forests brought men here just as the gold rush of 1849 did. The old method of logging with hand axe and saw led these loggers to believe they couldn't make a dent in the vast groves. One hundred and fifty years later we can look back and see the error of their ways. The loss to our pines, oaks, and redwoods is clearly visible; the trees that once covered forty-

LEFT: ***Santa Rosa Plateau.*** *The fog is coming over the rise, the sun is setting, and standing watch as it has for centuries is the Englemann oak. Much of its historic habitat has been lost and not enough young trees are growing to replace the old and dying ones.*

eight percent of the state are now reduced to only eight percent of their historic range.

Some of the forests first seen by Jedediah Smith in 1828 still dot the landscape. The term "old growth forest" reflects a current change in man's perception of the forest heritage still remaining. Whether it's our coniferous forests—pines and redwoods—or deciduous forests—oaks and sycamores—all are old growth, irreplaceable once they vanish. They are truly sentinels for all time!

Santa Rosa Plateau. *With the native bunch grasses turning brown in the warm summer sun, the evergreen Englemann oak stands as if raising its boughs to the sun. On the plateau where the oaks are protected, they grow to massive girth, providing shade and adding to the watershed on which species such as the western pond turtle, arroyo toad, and red-legged frog depend.*

Englemann Oak

This little-known oak has few romantic poems or legends surrounding its life. It is not the tallest, fattest, greenest, or grandest of California's oaks. It could gain a recognition, however, that none of California's other oaks have—the first listed as endangered. The Englemann now occupies only twenty percent of its historic range, mostly in San Diego County. The oaks are not found in the warm, dry coastal plain and do not grow above an elevation of 4,200 feet. They are often associated with the endangered coastal sage scrub, growing next to or intermixed with it.

Palomar Mountain, Black Mountain, and Mesa Grande near Ramona are a few of the Englemann's last strongholds, but its showcase is at Santa Rosa Plateau. The Nature Conservancy, along with dedicated individuals, had the foresight to buy 6,900 acres of Englemann habitat, saving it from extinction. What is it about the Englemann, or any oak woodland community, that should cause us alarm if it is lost?

The plant community on the plateau gives us a hint of the interrelationships. In the grassland savannahs amongst the Englemann are some of southern California's last vernal pools. They are seasonal wetlands during the winter for migrating waterfowl. In the spring, they turn to rings of color as cycles of wildflowers take the place of the receding water. These moist grasslands are among the last refuges for the endangered San Diego button-celery, thread-leaved brodiaea, and Parish's meadowfoam.

The creek bed of the plateau has tenajas (deep holes that hold water after the creek has dried) which provide water for the wildlife. Two species being considered for listing, the red-legged frog and western pond turtle, find refuge here. Mountain lions, bobcats, California quail, and thrashers all depend on the water in this ecological niche.

It is so difficult to make the point that a plant species as large as an oak can be endangered, especially when one can see what appears to be miles of it. The Englemann will end up being a controversial species, just like the spotted owl or California gnatcatcher. Its valuable habitat carries one of the highest price tags, as developers set their hopes on converting these woodlands into homes and businesses. The Englemann oak habitat can also be where Californians finally take a stand and say enough is enough!

Santa Rosa Plateau. *Amongst the oaks in the fields of grasses, an arroyo provides one of the last homes for the San Diego button-celery. Known to some as coyote bush, this endangered plant, like many others, has lost its fragile habitat.*

Murrieta. *Just below the Santa Rosa Plateau in an area barren of any civilization, the Stephens' kangaroo rat lives in the coastal sage scrub below the Englemann oak. The Stephens' kangaroo rat was first identified and described at Murrieta, and this is now one of the first places where it might become extinct.*

Santa Rosa Plateau. *In the boughs of an Englemann oak, high above the hustle and bustle of the forest floor, a great horned owl tries to stay concealed in its day roost. The largest owl in North America is vital to the health of the forest, as it keeps in check the many nocturnal small mammals that scurry about.*

Santa Rosa Plateau. *No one bird signals the coming of spring to the oak forest like the western bluebird. The bright flash of blue in the forest of green is a sure sign that nesting has begun. Bluebirds take advantage of natural as well as woodpecker-made cavities in the oaks to lay their eggs and raise their young.*

Santa Rosa Plateau. *Though not yet on the endangered species list, the arroyo toad is currently under consideration. The toad and its neighbors, the western pond turtle and red-legged frog, have been hit hard by the loss of their specialized stream habitat.*

Santa Rosa Plateau. *California's only native turtle, the western pond turtle, will soon be its first and only endangered turtle. Seeking out the waters in the spring to mate and raise their young, the turtles go underground in the fall and sleep through the winter.*

Central Valley. *Finding a dense valley oak forest is very difficult now, but once you do, your imagination can easily transport you back 150 years. Under the oaks would be mule deer and grizzly bear eating acorns, while high overhead the calls of thousands of birds could be heard.*

Central Valley. *The common western gray squirrel's role in the oak forest is vital. The squirrels eat much of the valley oak's production of acorns, but they also stash them, and then the acorns grow into new young trees.*

VALLEY OAK

But scattered over these hills and in these valleys are trees every few rods-great oaks, often of immense size, ten, twelve, eighteen, and more feet in circumference, but not high; their wide-spreading branches making head is often over a hundred feet in diameter of the deepest green foliage...

William H. Brewer, 1861

Like the passenger pigeon that once blanketed the skies of the east, so did the valley oak cover California's landscape. Accounts of mile-wide forests covering the valley floor and giant oak specimens living amongst them were legends that traveled the world. One oak in the Central Valley brought a botanist from England just to see its majestic magnificence. With John Muir and a company of others, he searched for and found the tree, which was named the Hooker Oak in his honor.

Most Californians are familiar with the valley oak. It has not long vanished like the grizzly bear which once fed on its acorns. But surprisingly, this symbol for many of California's towns and businesses is in real trouble.

The valley oak was quickly recognized by the farmers who came west in search of gold as occupying the best land. The oaks fell at a precipitous rate, first by axe, with a horse team to pull up the stumps, then later by explosives, which took care of the entire job with one blast.

The oak's importance to its ecosystem can easily be witnessed in an afternoon. High in its boughs, the acorn woodpecker, northern flicker, scrub jay, and yellow-billed magpie are just a few of the birds flitting about. In its canopy, looking over the grasslands, might be a hawk—red-tailed, red-shouldered, or Swainson's—or a golden eagle. The raptors look for the unsuspecting western gray squirrel busily caching some newly fallen acorns. The grizzly bear once made a fall feast of the tons of acorns the oaks produced; the black bear still does. The mule deer also dines on the acorn harvest while the coyote waits in the shadows to take advantage of any lack of vigilance.

Spacious forests of valley oak are hard to find in the Central Valley. Majestic individuals still dot the landscape, and little groves can be found covering a hill. Reflections of the great stands can be found at Caswell State Memorial Park. Though greatly reduced, this stand is still spectacular. This is the last refuge for the riparian brush rabbit, which numbers less

Central Valley. *In the boughs of every valley oak, as far as you can see, are band-tailed pigeons. It is fall and the oaks are full of acorns, the pigeons' favorite food. They can strip a tree bare during their feast, knocking down as many as they eat so that other wildlife can partake in the bounty.*

than one hundred, and is an important area for the threatened Swainson's hawk.

In its statuesque winter shape, dressed out in spring greens, or lighting up the heavens with its red fall attire, the valley oak is magnificent. Loss of habitat is only part of the reason the valley oak may be in trouble. The tree's own mechanism for regenerating is the other. Oaks produce thousands of acorns because not many actually ever grow into trees. Whether eaten as acorns or as new spring saplings, very few become large oaks. Though the oak is not listed as an endangered species, this noble California symbol could become like the California grizzly bear, seen only on the state flag.

ABOVE: ***Lake Arrowhead.*** *The lodgepole chipmunk is a busy little creature, darting about the forest floor and trees. It serves a vital role in planting seeds from the plants it eats, and also in becoming prey for many of the forest's predators.*

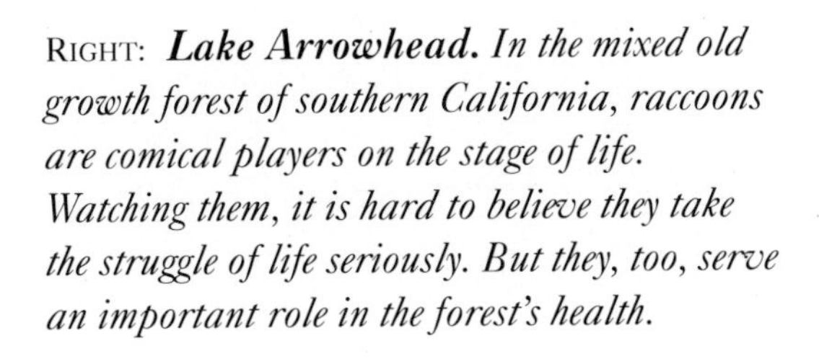

RIGHT: ***Lake Arrowhead.*** *In the mixed old growth forest of southern California, raccoons are comical players on the stage of life. Watching them, it is hard to believe they take the struggle of life seriously. But they, too, serve an important role in the forest's health.*

Southern Old Growth

The vision that comes to mind for "old growth" is tall redwoods and fog-shrouded hills, not the forests of southern California, but old growth is defined as trees over two hundred years old, a definition many of southern California's forests fit. Although they might not have the romantic allure of their northwestern cousins, they are just as valuable.

Seventy-four conifers are known in the western United States. Fifty-seven of those are found in California, twenty-seven of which are endemic. The list of trees includes some of the finest: Douglas-fir; digger, coulter, yellow, Jeffrey, and ponderosa pines; black, blue, coastal, or interior live oak. They constitute over four million acres in national forests alone; is this enough to safeguard them and their wildlife?

Old growth forests change their complexion according to where they have taken root. They vary by being on north- or south-facing slopes, or if they are located in the Santa Ynez, Santa Ana, San Bernardino, or San Gabriel mountains. The soil, water, amount of sun, and temperature are other variables.

These forests harbor a variety of wildlife under their canopies; probably the most spectacular are the birds. Woodpeckers are the most dependent on the trees. These lumbermen of the bird world help keep the forest healthy while obtaining their food; they drill in tree bark and flesh for insects that quite often harm the trees.

Many birds of the old growth forest are secretive, often considered rare, or are listed as species of special concern. The flammulated owl and black swift are two such birds. The flammulated owl is smaller than a dollar bill and inhabits abandoned woodpecker holes. The black swift is North America's fastest flying bird and nests behind waterfalls, supported by the forest's watershed and the insects it attracts. Both of these species are migratory and dependent on the forest habitat each spring to raise their families.

This is the home of the southern spotted owl, the cousin of the controversial northern spotted owl. The owls look alike and have the same biology, but the southern species

Lake Arrowhead. *The old growth forests of southern California support the southern spotted owl. A close cousin of the northern spotted owl, which is a threatened species, the southern spotted owl might face the same fate. This graceful silent keeper of the forest is threatened by development more than logging, but the loss of habitat still has the same effect.*

Prairie Creek. *The sun is blocked, the forest floor darkened by these towering giants. The majesty and solitude of the old growth redwood forest is unmatched.*

has slightly different habitat requirements. It inhabits old growth pine and oak forests as opposed to the dense redwood and fir old growth forests of northern California. For both owls, loss of habitat means loss of birds.

The importance of these forests, no matter where they are found, goes beyond the birds and mammals in them. The forests are key to man's survival. The role they play in producing and replacing oxygen in the atmosphere is crucial. Probably just as important is the watershed they protect. With southern California being a coastal desert plain, the snows that fall and are held in the forests are critical to water supplies.

The old growth forests of southern California are visited by millions of people every year. More and more are moving permanently to these forests to be close to nature. This appreciation is loving the forest to death. In some areas of the forest, acres upon acres of pines are brown, dead from strangulation. Smog has plugged the pores of the trees so they can't breathe.

"Millions of acres" gives most people the sense that the few trees they cut for this shopping center or that home won't make a difference. Probably just as endangered as northern California's forest, maybe even more so, this refuge for man and creature is by no means secure for future generations to enjoy.

Northern Old Growth

The magnificence of these towering giants is indescribable. The first written accounts in the 1800s were doubted until a slice of redwood was shipped back to the east coast. Except for a small extension into Oregon and an ancient cousin in China, the redwood is strictly a California phenomenon. These two-thousand-year-old "kids," scraping the sky at 360-plus feet, are being overshadowed, their fate resting on the remaining five percent not yet forested. And their controversial inhabitant—the northern spotted owl—has not made things any easier for them!

The coastal redwoods are often confused with their inland relative, the sequoias. The redwoods have the height while the sequoias have the girth, but both are endangered. Redwoods are restricted to the maritime fog belt of the coast, usually on the north-facing slope. Their tall size and large limbs collect the fog on flat needles until it forms a drop which then hits the ground. These shadowy, moist conditions aid more than the redwoods. The coastal old growth forest is also comprised of other trees, such as the Douglas-fir, an often overlooked species. Obtaining grand heights, growing in dense forests, it too is critical habitat for many species and the watershed. Oaks such as black, live, and Garry, along with madrona and California bay add to the canopy of the old growth forest.

On the forest floor numerous shrubs have evolved to tolerate and flourish in the dark, damp environment. Oregon grape, poison oak, and wax myrtle, along with sword, chain, and wood ferns, create a carpet of incredible thickness.

This old growth forest majesty can stand on its own, but those that call it home also add to its splendor. The Roosevelt elk roams the shadowy world of the redwoods, emerging occasionally in the meadows or on the beach. Overhead, watching from the bough of a redwood might be the northern spotted owl, a species whose plight has come to symbolize the old growth forest.

More mysterious yet is a little bird which is just reaching the spotlight. The marbled murrelet is a seabird which is the size of a robin. It spends approximately three-fourths of the year at sea, never making landfall. When it comes time to raise their young, the murrelets leave their watery world and head inland. In humid, virgin forests near their coastal feeding sites, they make their nests high, possibly hundreds of feet above the ground on a large limb covered with moss. This very specialized nester is now listed as threatened, as the sites it requires for nesting are restricted to only a few locations.

Looking at the cities of Monterey and Palo Alto, it is hard to envision that one hundred and fifty years ago, the hills were covered with redwoods. Riding a boat up the Klamath River, it is just as hard to accept the clear-cutting of entire hillsides, knowing what stood there before the axe fell. When the virgin or old growth forests are gone, an entire ecosystem will vanish. This point has been and will be argued until it is proved by cutting what is left. The blinders of the short term are firmly in place on many and without the general public taking a stand on protecting our natural heritage, these giants will be threatened with becoming decks and fences.

SPOTTED OWL

No larger than a football, the northern spotted owl has turned out to be David to the logging Goliath in the 1990s. The heated controversy over listing the spotted owl has completely overshadowed just how special this owl really is. On a walk through its habitat, you can pass right under it and it will not make a sound or ruffle a feather. As it glides through the forest, the air is barely disturbed by its passing. For all its silence, it amazingly has at least fourteen "hoots" to communicate with its mate or young. The owls mate for life, raise two to three young each spring, and spend much of their life in the same section of forest. They prey on woodrats, flying squirrels, and small birds. The owls are mostly nocturnal, and roost on tree limbs during the day.

Their only mistake was evolving to depend on old growth forest. Of course that fact is in dispute, just as everything relating to the owl's habitat is. One side of the argument advocates saving the remaining five percent of old growth, while the other says to cut it because the loggers need jobs. Once the forest is gone the loggers will certainly lose their jobs, and in the meantime the owl will disappear.

Prairie Creek. *The Roosevelt elk are a marvel as they move around the old growth forest. Whether up in the midst of the forests in the grass meadows, or foraging here at the beach, their majestic beauty only heightens one's awareness of the wonder of the old growth forest.*

Coastal Sage Scrub

California's Most Endangered Habitat

The coastal sage scrub communities are home to approximately fifty species of plants and animals presently considered threatened or endangered (two of these are probably extinct; they were never listed). This is directly related to massive habitat loss and degradation. This vastly misunderstood, drab gray carpet of sagebrush is a critical piece of California's landscape puzzle.

The entire worldwide distribution of coastal sage scrub is here; it extends from the San Francisco Bay region to Baja California. The composition of plant species changes from the north to the south as climatic conditions become drier, so the scrub is divided into two types—northern and southern—with Santa Barbara as the arbitrary dividing line.

The plant species that give the coastal sage scrub its character are the California sagebrush and California buckwheat. Manzanita grows only in the northern coastal sage scrub and prickly pear cactus is found only in the south. The southern coastal sage scrub extends further inland than the northern, but both have suffered the same devastating losses from land conversion.

LEFT: ***San Pasqual battlefield.*** *The coastal sage scrub does not have overwhelmingly apparent beauty, nor does it appear to support much life. The cactus conveys the impression of harshness and longevity. But this appearance is misleading; the coastal sage scrub is a habitat of quiet beauty that supports a wealth of life.*

San Diego County

This lonely habitat was ignored for decades, but is now very much at the front of the stage. Once covering 2.5 million acres from northern Ventura County to Baja California, the southern coastal sage scrub has been reduced in just the last thirty years to only 250,000 acres. This horrendous and tragic loss has made the coastal sage scrub the most endangered habitat in the world!

This ecosystem is characterized by a diversity and variability that resists being pigeonholed. Three geographic associations have been identified within the southern coastal sage scrub—Venturan, Riversidian, and Diegan. Of course, these do not come with boundary lines; their association reflects a north-south and coastal-inland gradient of increasingly drier conditions. The scrub's character is also affected by other factors, such as changes in elevation, fire history, and whether the terrain faces north or south.

The plants that make up the sage scrub are a scraggly gathering of shrubs, characterized by sagebrush and prickly pear cactus. Other cactus, including the rare coast barrel cactus, and common plants such as buckwheat also comprise the sage scrub community. Sagebrush and prickly pear are visually the most dominant plants of the ecosystem. Their edge effect is crucial to many species, such as the coastal cactus wren and California gnatcatcher.

There are a number of indicator species telling us that the coastal sage scrub is endangered; two are birds, the very controversial California gnatcatcher and the overshadowed coastal cactus wren. The coastal cactus wren eats only insects that feed on the prickly pear. These same cacti support the wrens' bulky nests, which are used not only to raise young but also for year-round nighttime roosts. For these reasons, it is very likely the coastal cactus wren is more threatened with extinction than the gnatcatcher. Because the coastal cactus wren has not yet been recognized as a distinct subspecies, listing under the endangered species act is very difficult, so legal protection will be an uphill battle.

The orange-throated whiptail lizard is another colorful species of the coastal sage scrub. While most lizards lie in wait to ambush their prey, the orange-throated whiptail goes in search of its prey. Its flashy color makes its two- to three-inch body with its six-inch tail noticeable as it scurries through the scrub. The lizard's populations are scattered, so fragmenting the coastal sage scrub critically affects it.

The magnificent mountain lion still exists in southern California. The largest of our remaining predators depends on important corridors associated with coastal sage scrub; males travel from island to island of fragmented habitat to find mates. These corridors are already being lost at an alarming rate, severing important gene pools among these large cats.

On the other end of the mammal spectrum is the Stephens' kangaroo rat. This small, endangered species lives on the fringe of the coastal sage scrub, mostly on plateaus or in flat regions, and makes use of the open areas within the scrub. Its endangered status affords the coastal sage scrub it inhabits some protection under the endangered species act, but this does not generally include the same habitat the California gnatcatcher occupies. The Stephens' kangaroo rat is the most recently listed of the five endangered kangaroo rats in California, and has the best chance of escaping extinction.

The loss of coastal sage scrub affects all things great and small; the Quino checkerspot butterfly is a good example. One of southern California's vanishing butterflies, it is currently a candidate for listing.

Many plants of the coastal sage scrub are already listed as endangered; others wait to be. One of the more unusual ones to be listed is the short-leaved dudleya, a specialized inhabitant of the coastal sage scrub. This two-inch-tall succulent lives in shallow sand pans, where it simulates the appearance of a small red pebble when it first emerges. It has lost fifty percent of its habitat in the last four years; proposed development will consume its largest population in the near future and no protection plan is being considered. With current trends the dudleya most likely will become extinct.

One aspect that is often overlooked is the importance of the coastal sage scrub on species that are not inhabitants of it. The best examples are the birds of the Santa Margarita River riparian track. These birds, including the least Bell's vireo, often forage in the coastal sage scrub bordering their riparian home. The loss of any habitat affects not only its own inhabitants, but also those that live in the neighboring habitats.

"The most endangered habitat in the world" is a status no place should bear. Neither an inspirational landscape nor a home to glamorous species, the coastal sage scrub nevertheless is a vital and unique ecosystem that cannot be replaced. There is a very real possibility that it may not exist by the turn of the century. The fight over natural heritage versus money is a classic, and the newest habitat to see this fight is the coastal sage scrub. The question is, will the ending be written as a tragedy?

California Gnatcatcher

The most infamous inhabitant of the coastal sage scrub is the California gnatcatcher. This small, gray, four-inch bird weighs only six grams and numbers less than two thousand pairs. Populations that once inhabited Ventura and San Bernardino counties are now extinct, and its last refuges in Orange and San Diego counties are under direct fire. A non-migratory, sedentary bird, under normal biological conditions the gnatcatcher never leaves the coastal sage scrub. Its home range can vary in size from two to fourteen acres; a bird normally inhabits the same plot of land its entire life. Nesting duties are shared by both sexes.

Like the vireo, the gnatcatcher is taking all the heat for listing since it will stop the greatest amount of development. Bulldozers have been clearing hundreds of acres, trying to beat the listing of the gnatcatcher. There is a saying that the most endearing thing about the six-gram gnatcatcher is that it can stop a bulldozer; only time will tell if that is true. The gnatcatcher is known to inhabit only 54,000 of the remaining 250,000 acres of coastal sage scrub. Long-term survival for both the gnatcatcher and the habitat will require more than a token population here and there.

San Bruno Mountain. *The coastal sage scrub of northern California appears quite different from its southern cousin. Foggy and windswept, often right on the coast, it is beaten down but usually green. In spring the flowers vital to the butterflies of San Bruno Mountain bloom, painting the hillsides with color.*

San Bruno Mountain

Rising to a height of 1,314 feet, San Bruno Mountain is an island of open space in a sea of urbanization. San Bruno Mountain State Park south of San Francisco attracts thousands to its natural wonders. San Bruno Mountain does not have tall peaks, deep dark forests, or rivers of foaming white water, but it is one of the last principal communities of northern coastal sage scrub. Except in those spaces dedicated to the comforts of man and his car, the mountain is covered in coastal sage scrub, with small pockets of remnant oak woodland, chaparral, and riparian habitat.

Because it is so close to the coast and to San Francisco Bay, the climate on the 2,326-acre mountain is not typical for coastal sage scrub. In the summer it receives blanketing fog, and in winter storms cover its heights; both bring valuable moisture. It averages twenty to twenty-five inches of rain a year, and temperatures rarely reach freezing or rise above eighty degrees. This moist, mild climate encourages even the smallest plant to grow into a robust specimen. It is not hard to understand why San Bruno Mountain supports such rich plant life.

Historically, the mountain and adjacent lands were host to six distinct plant communities; now only two remain. Research shows that thirty-three percent of the flora is endemic to California, twenty-two percent to the Pacific coast, and nine percent to North America. These communities are home to 384 native plants; of these, at least 14 are considered rare or endangered.

Manzanita has been particularly hard hit in the northern coastal sage scrub. The San Bruno Mountain manzanita is a low evergreen shrub; only four populations of it still remain. This endangered plant depends on the sandstone outcrops in the coastal sage scrub community. The Pacific manzanita is also endangered. Slightly different from the San Bruno, it seems to be suffering from fire prevention policies; it needs fires to produce new young plants. The Montana manzanita, though not listed, is considered rare and will eventually join its cousins on the endangered species list. They all have suffered from development; their fight for survival depends on the preservation of San Bruno Mountain.

Many of the smaller plants, including some wildflowers, are also rare or endangered. The coast rock cress, Franciscan wallflower, San Francisco owl's clover, and San Francisco campion are just a few of the specialized inhabitants that are not faring very well. These plants depend on insects to pollinate them so they can reproduce, but with so many plants lost or becoming lost, a group of these pollinators has been adversely affected.

The San Bruno elfin and mission blue butterflies are endangered species hanging on in this island of coastal sage scrub. Larvae of the San Bruno elfin butterfly depend on stonecrop, a plant that can only be found in this area. The mission blue butterfly larvae depends on a species of lupine which grows on this mountain. Other butterflies inhabiting the mountain, though not endangered, depend on the same lupine. The callippe silverspot butterfly, bay checkerspot butterfly, and San Francisco tree lupine moth may soon join the others on the endangered species list.

San Bruno Mountain's story is typical of most of California's endangered habitats. With the loss of such a high proportion of its historic range, those species that evolved to depend on it become endangered. The narrow view would say that it is just a few plants and a couple of butterflies, things that are common to all gardens. This is a dangerous and destructive view. The lessons of San Bruno Mountain point this out; a steady decline has occurred as the habitat has been removed. Yes, it might only be plants and butterflies now, but later...? San Bruno Mountain itself is not fully protected; development reaches towards every possible square inch.

San Pasqual battlefield. *Overshadowed by the California gnatcatcher, the coastal cactus wren is just as endangered. This amazing bird makes its nest right in the cactus; this female is on her way into the nest with a mouthful of nesting material. Notice how masterfully she perches on the cactus spine.*

San Bruno Mountain. *On a lonely ridge line high atop the mountain, the unique flowers that support the San Bruno elfin bloom. In the natural world, it is always amazing how small creatures such as butterflies can evolve to be so specialized that they depend on one flower on one ridge.*

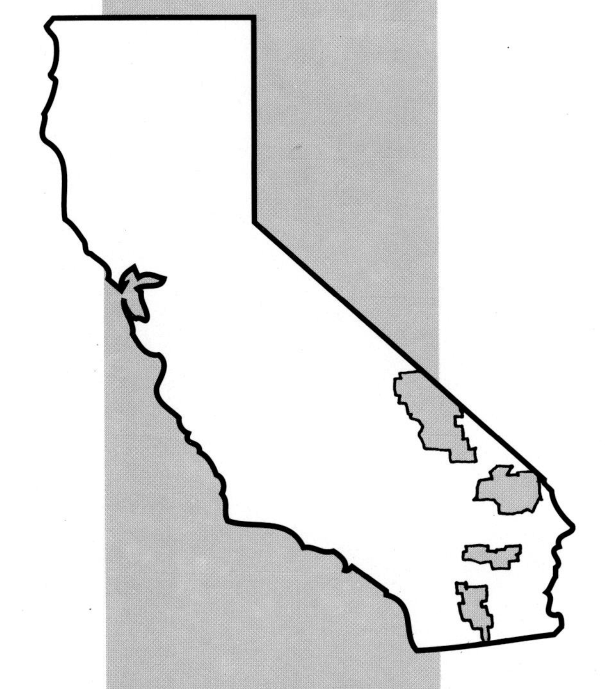

The Desert

A Gem in the Rough

There is something attractive in the very name of the desert. It is invested in our minds with the stories learned in childhood, of its wonderful wide stretches of sandy wastes, its mirages, and its caravans, all of which have been so generously adorned with the splendors of diction, and dressed out in the gorgeous robes of imagination.

James Capen Adams (Grizzly Adams), 1854

As a group, California's three deserts —the Great Basin, Mohave, and Colorado—have a greater biodiversity than the twelve other deserts of the world.

By definition, a desert is an arid area with insufficient available water for dense plant growth; more precisely, an area of very sparse vegetation in which the plants are typically solitary and separated from each other with a large proportion of barren ground. The California desert is a region of environmental extremes, with less than ten inches annual rainfall, high evaporation, low humidity, and a high percentage of nocturnal animals.

Each desert merges into its neighbor and only in some areas can any line be drawn between two deserts. They are best separated by biological rather than topographical factors. Mammals, birds, insects, and reptiles move freely between deserts as food, water, or other factors dictate. The plants are normally restricted to living in just one desert habitat.

LEFT: ***China Ranch.*** *The barren loneliness of the Colorado Desert, the mountains of a thousand colors, and the timeless sands add to its mystery. Such places as this are rarely visited, so life can be seen as if time has held still for centuries.*

Colorado Desert. *These thin, sharp, barbed spines catch the morning light and transform it into sun rays. The bigelow cholla, or jumping teddy bear, is a cactus that is found only in the Colorado Desert.*

Colorado Desert. *In the spring when its sharp tangle of spines gives way to flowers, the Simpson's hedgehog cactus truly stands apart from its thorny neighbors.*

Colorado Desert

Now, I wish I could describe this desert so that you might really appreciate what it is-a great plain, rising gradually to the mountains on each side, sandy, but with clay enough in the sand to keep most of it firm, and covered with a scanty and scattered shrubby vegetation.

William H. Brewer, 1863

The Colorado Desert is by far the most barren of the three deserts in appearance. Plants such as the scraggly creosote characterize its landscape. It has points below sea level, such as the Salton Sea, and the New York Mountains rise to nearly 8,000 feet. But generally, the Colorado Desert is less than 2,000 feet in elevation. The desert is bordered by the Coast Range on the north and the Peninsular Range on the east; the Transverse Range divides it from the Mohave Desert. These mountains are the reason for the desert's existence; they stop the moist Pacific storm clouds from reaching its landscape. The Colorado Desert averages only one to five inches of rain a year.

The Colorado Desert is California's portion of the Sonoran Desert; it has extremely mild winters and blistering hot summers. This seasonal variation prevents many of the plants associated with this desert from growing in the neighboring deserts. The Mohave Desert's winters are colder, creating an invisible wall that keeps out these heat-dependent southern plants.

The most distinctive plant of the Colorado Desert is the creosote bush; this successful plant covers vast areas. After the winter rains, its yellow blossoms burst onto the scene for a short time. Following this are its fuzzy white seed balls, giving the desert the appearance of a light dusting of snow. Most of the year the creosote glistens in the sun as the wind hammers at it. The glistening effect comes from its "varnished" leaves, which reduce evaporation from the sun and wind.

These thousands of square miles of creosote provide a stable base for other plants. One is the night-blooming cereus, hiding under the creosote while growing two to five feet in height. Known also as the "queen of the night," its blossoms appear at night and are described as wearing halos. Underground, its beetlike root is a moisture-storing organ, attaining a size of five to eighty-five pounds!

Another specialized plant family in the desert is the cactus. The jumping teddy bear that inhabits the Colorado Desert is one of the more spectacular. The bigelow

FRINGED-TOED LIZARD

This endangered species is a specialized marvel, residing in a specialized habitat. Known by old time residents as the "sand swimmer," the Coachella Valley fringed-toed lizard lives exclusively in blow-sand fields. To avoid its predators—greater roadrunner, whipsnakes, loggerhead shrikes—and the heat of the desert, the fringed-toed lizard "swims" below the surface of the sand to escape. It has evolved in every way to take advantage of its habitat. It has a sharp-edged jaw and chisel-shaped skull for "diving" into the dunes. Its scales are smooth and round to reduce friction, and "fringes" on its hind toes provide traction as it "swims." It has even adapted to breathe under the surface without inhaling the sand. In November, like many reptiles, the fringed-toed lizard simply buries itself in the sand and slumbers until the warmth of the February sun. In May the males take to courting, performing what appear to be push-ups to attract mates. The female's eggs hatch after four to eight weeks of incubation. The only problem the fringed-toed lizard has not evolved to solve is that of living in a habitat desirable for resort condos. Only a small fraction of its habitat is protected, and that which has not already been developed faces extinction.

Colorado Desert. *The black-tailed jackrabbit is no stranger to the Californian, but this very successful mammal is still an amazing animal. It thrives in an environment where man takes to his air-conditioned car and brings water with him. This creature finds all it needs—food, water, and shade—among the cactus and in the sun. Even the dry, waxy creosote and its bright yellow blossoms nourish this female, who is getting ready to have her litter.*

Kelso Dunes. The windswept artistry that molded and carved these sands is truly one of California's great wonders. No matter where you turn to get a glimpse of nature's handiwork, the depth, loneliness, and majesty overwhelm and delight the senses.

cholla (its proper name) grows amongst the creosote and can amass into large gardens, such as at Pinto Basin. The seeds are infertile, so the cactus propagates by means of small sections of the parent plant that break off. The barbed spines of these sections attach to anything getting too close, like a coyote's tail, and thus are transported to new areas.

Surprisingly, these jumping teddy bears provide homes for wildlife. The desert woodrat, most commonly known as the packrat, makes "nests" with masses of this thorny cactus. The woodrat creates runways within a fortress of barbs to protect itself from the hunting coyote. Unlike the kangaroo rats that inhabit the same desert, the woodrat depends on green plant material to obtain moisture, and is unable to obtain it from seeds alone. The woodrat can eat cacti and plants that are toxic to other animals. Wondering how it beats the heat? That house of spines lets the breeze in while blocking out the sun. The desert speedster, the greater roadrunner, also uses the cholla for a home. It builds its nest right in the cactus, dodging the spines to enter or exit. The cactus wren, whose name is derived from its thorny home, uses the same technique.

The desert sands have long been a familiar symbol for their lack of water. The Colorado Desert contains many sand dunes, including the largest in California, Kelso Dunes. This amazing chronicler of time, which seems to contain no life, is one of the

Left: ***Stovepipe Wells.*** *The story of the passage of a greater roadrunner has been left for those who look down and read what the sands have to say. Our minds can fill in the gaps as we imagine seeing the roadrunner, with a newly caught lizard in its bill, darting across the sand to deliver it to its waiting young at the nest.*

Right: ***Tumco Wash.*** *There is nothing as sweet, soft, or gentle as the California leaf-nosed bat. Unlike many bats, the California leaf-nosed bat does not like to hang touching his neighbor even though they share the same day roost. At night the bats leave their roost in grand fashion and venture out over the desert to find and consume thousands of insects.*

most fragile and endangered ecotopes, or habitat types, of the Colorado Desert.

The story of the dunes is best told in its sands. An early morning walk reveals the tracks of hundreds of nighttime inhabitants eking out their existence. Two parallel lines with a third line running through the back center tells of where a kangaroo rat landed. Where two of these marks come together in a jumble there was a territorial fight between two kangaroo rats. Where one just ends next to long, parallel tracks, a rattlesnake caught its evening meal.

A number of small mountain ranges are nestled in the Colorado Desert, each unique because of its location, such as the Cargo Muchacho Mountains. Shafts sunk for gold mining at the turn of the century replaced the natural crevices where many species of bats lived. California leaf-nosed bats come to the shafts because of the exposed geothermal pockets, which provide an eighty degree room of high humidity. After foraging at night for insects, the warmth is important for these warm-blooded animals.

Researching bats requires descending into the pitch-black mines. These bats want nothing to do with humans, avoiding the light of our flashlights, leaving rooms where our voices are too loud. Hanging on the ceiling deep in the mine, they wait for night to venture out in large flights and feed on insects in the surrounding desert. These small creatures are able to eat pounds of insects each night, and, as a collective colony, are a tremendous natural form of pest

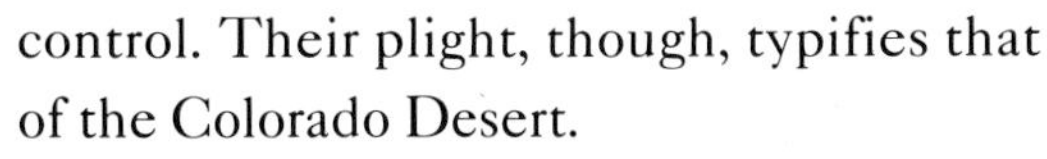

control. Their plight, though, typifies that of the Colorado Desert.

Like the California leaf-nosed bat, the majority of the Colorado desert life is hidden, much of its habitat already lost or degraded. And with it have gone many species, some unknown to science. This is a hostile environment, but it is also a fragile one. The unique conditions that created the desert arose over millenniums; damage in our lifetime might just undo what time seems to have forgotten. The future for the Colorado Desert and many of its inhabitants is in jeopardy. Many are waiting to be listed to obtain protection, but none are able to withstand the punishment of current times.

Joshua Tree. *Probably the best known of all desert dwellers, the greater roadrunner is uniquely adapted to the rigors of this harsh environment. One of its adaptations is deriving the moisture it requires from the prey it eats.*

Joshua Tree. *The Joshua tree is a mysterious tree indeed, growing in a million different forms and as many directions. Provider of food and shelter, the Joshua tree is as at home in the desert sun as in the winter snows, which are as white as its spring blossoms.*

JOSHUA TREE

Joshua Tree National Monument offers a rare look at the merging of two deserts, the Mohave and Colorado. There's no fence separating the two, though a sign is present to make the unobservant note the passage from one desert into another. The monument runs on a northwest heading with the Colorado Desert at its southern end, the Mohave at the northern. The monument's 560,000-acre landscape disguises its gentle upslope with rock formations, hills, and mountains. Entering the monument from Cottonwood Spring and driving up Pinto Basin, the transition zone is on the horizon. Cottonwood Spring is the lowest elevation in the monument where you can find the ocotillo.

The ocotillo's long, spindly limbs give it an upside-down appearance, as if its roots were dangling in the air. Most of the year its tawny-spined stems wave in the breeze, earning it the name "coachwhip." The rains change the ocotillo's whips from brown to green; dangling like pennants on their tips are its bright red blossoms. Because of its thorny stems it is considered a cactus, but actually the ocotillo is more closely related to the primrose and olive family.

Further up Pinto Basin, the creosote and ocotillo begin sharing their habitat with the bigelow cholla, diamond cholla, barrel, hedgehog, and mound cactus. This is the northernmost reach for most cactus, as the freezing winter of the Mohave Desert creates an impenetrable boundary.

Symbol of the Mohave Desert, the Joshua tree signals the change in deserts. Occurring between 2,500 and 5,000 feet in elevation, it is believed to need the frost and occasional snows of the Mohave Desert for proper growth. Trees of a different standard, many desert flora books refer to them as tree lilies because of their bunches of white blossoms covering the branch tips in spring.

The Joshua tree must withstand the harshness of the summer heat, the cold frost of winter, and long periods of drought. Its tough, leathery, spiked leaves are shorter than other yuccas (its cousins), preventing evaporation in the summer and frost damage in the winter. These giants of the desert support a vast and diverse community of creatures, which usually spend their entire lives within a single tree. One inhabitant is a moth.

The symbiotic relationship between the tegeticula moth and the Joshua tree is a classic. During its evening rounds, this moth collects enough pollen to make a small ball. With this, the female moth lays her eggs in the ovary of the Joshua flower and pollinates the Joshua tree with the pollen ball. By the

Joshua Tree. *The desert dandelion and Wallace's woolly daisy paint the desert floor in a carpet of gold. The boulders frame in their color, which otherwise would explode across the horizon.*

Fortynine Palms Oasis. *The Nelson's or desert bighorn sheep sneaks down to the oasis for a cool drink. Before any human visitors have a chance to find him at the oasis, he effortlessly walks back up the vertical walls where, from his lofty perch, he can safely observe.*

time the moth larvae hatch, the Joshua ovary has produced seeds that the larvae consume; enough seeds are left to generate a new Joshua tree.

Joshua trees also serve the desert birds. The Scott's orioles which return to the desert each spring depend on the blossoms. They also use the Joshua tree for secure locations for their hanging nests. A ladder-backed woodpecker chisels out a home in the Joshua's trunk, and it later provides a nesting cavity for a western screech owl. Great horned owls perch on the Joshua's limbs in the dark of night, making a meal of any unfortunate desert woodrat, kangaroo rat, or mouse living in the debris of the tree.

The monument is home to four oases: Twentynine Palms, Fortynine Palms, Cottonwood Spring, and Lost Palms Oasis. The centerpiece of these oases is the native fan palm. These oases are just as you might picture them, with springs bubbling up from the earth, pools reflecting the swaying palms, and the sounds of birds coming from all directions.

The fan palms produce dates which mature during the summer and ripen in the fall. When hanging in the tree, the fruit provides food for many different bird species. When the dates dry and fall to the ground, many of the desert's nocturnal mammals come to gather the harvest. The coyote is one of the biggest fans of this tough fruit, eating it whole, later spreading the seeds through his scat to produce more palms as far as fifty miles away.

Though only a remnant now exists of the magnificent forests of pinyon pine and oak of a hundred years ago, Joshua Tree is still a natural monument. The monument, and all that call it home, is protected from the ravages of current development. But outside the monument development continues, destroying habitat and gobbling up precious water resources. Many pools and springs of decades ago have already retreated underground. Some of the greatest battles over the environment are shaping up to take place over the desert. In the near future, the fate of what once was thought of as an inhospitable environment will be decided. Will the desert become literally a wasteland as many already believe it is, or an oasis of our commitment to protect the land we inherited?

Fortynine Palms Oasis. *Water rises from springs to nourish the native palms in the oasis, which provides critical water and shade for the desert's inhabitants.*

MOHAVE DESERT

This is a country of three seasons. From June on to November it lies hot, still, and unbearable, sick with violent unrelenting storms; then on until April, chill, quiescent, drinking its scant rain and scanter snows; from April to the hot season again, blossoming, radiant, and seductive.

Mary Austin, 1903

The Mohave Desert is one of the most diverse deserts in the world. Most of the Mohave lies between 2,000 and 4,000 feet, and serves as the transitional zone for many plants between the Colorado and Great Basin deserts. It runs south of Olancha on a jagged line to Palm Springs, then east to Nevada. This desert, which freezes eighty nights out of the year and receives one to five inches of rain a year, is uniquely Californian.

This land of little rain still bears the impressions of man's early inhabitants. The pictographs of the Paiute Indians, such as those at Fossil Falls, are a testament to their life and that of the wildlife and plants they celebrated. Ruts made 130 years ago by the pioneers' covered wagons are still visible, carved in the hard clay soil of the Mohave.

The creosote bush is another sign of the Mohave's immortality. This pungent-smelling, unglamorous, and dull-colored plant will never make it in any gardening magazine. But it can live to be ten thousand years old, and one group of creosote in the Mohave is thought to be the oldest living thing in the world!

Heading north on Highway 395 from the town of Mohave, a canyon of enchantment greets the traveler. Red Rock Canyon's red clay cliffs, white sand formations, and water-etched sculptures are a showcase of Mohave Desert habitat. The washes, running wild with the winter rains, carved the canyon's unique architecture long ago. They still dominate the flat lands of the canyon and the plants and wildlife that call them home. The Joshua tree is sparse, and even the creosote has a difficult time taking hold. The same is true for the red rock tarplant, a rare plant up for federal listing as endangered. Its habitat is in the canyon's ephemeral streams and washes, and the recent years of drought reduced its numbers by ninety-five percent.

Rising out of Red Rock Canyon, the traveler sees a new realm of the Mohave Desert; the Joshua tree disappears and the landscape becomes almost bleak. Creosote forms the only canopy available to plants, protecting the many belly rubbers that inhabit this regions. Many of these two-inch plants are threatened or endangered because of sheep and cattle ranching, as well as off-road vehicles.

Tucked into secret corners of the Mohave Desert are special treasures; the grandest is Death Valley National Monument. This celebrated landscape is home to many examples of the Mohave's richness: marshes, canyons, oases, and rivers. A number of pupfish species have evolved to specialize in living in certain springs. Plants have taken hold and evolved to certain soil alkalinity levels. And one of the most magnificent inhabitants of the Mohave has a stronghold in Death Valley, the desert bighorn.

Known as the Nelson's bighorn or mountain sheep, it has long inhabited the desert. The only one of three subspecies in California not threatened, the Nelson's lives in the high ranges of the Mohave, moving through the hills as if they were super highways. Specialized hooves provide a sure grip even when running up sheer, jumbled rock faces. Dependent on surface water, the bighorn makes pilgrimages almost daily on ancient trails to springs. Today, a dedicated group conducts constant programs to improve or create new watering holes as

Mohave Desert. *The grasshopper mouse is the only carnivorous mouse in California. It preys upon insects including grasshoppers. It is also unique because it bays at the moon in a way. Rising on its back legs and pointing its head towards the sky, it emits a high frequency, single note. This call is so high-pitched that many people are unable to hear it.*

Desert Tortoise

In spring, when the rains have set the sands ablaze with wildflowers, a relic from a bygone era emerges. The threatened desert tortoise lumbers out of its burrow and forages, just as it has for more than two million years.

The tortoise ranges from the northern edge of the Mohave Desert south into the Colorado Desert. The heaviest of southwest reptiles seems an unlikely endangered species, especially since it lives in a "barren" region of the state and spends much of its time underground.

The tortoise's plight is due to a condition becoming more and more common — fragmented habitat. Highways and roads, development and recreation have created "islands" of habitat, isolating many of the tortoises. Further destruction of the habitat caves in its burrows, crushing the tortoise when it's inside or destroying its shelter from the summer heat. The tortoise is as controversial as any endangered species in the country; only public awareness will ensure it a safe home in our desert.

Red Rock Canyon. *Water-carved, windswept, and defined by the ages, this rock canyon is a unique California landscape. As if the red rock, soil, and sage are not enough, a rare alpenglow fills the scene to intensify the drama of the desert.*

those ancient ones disappear because of lowering water tables.

Whether driving to the Sierra or to Death Valley, you pass one memorial to progress. The white salt flats of Owens Lake are the only reminder of this once vast, rich, body of water. Turning to dust with water diversions, Owens Lake can no longer support the thousands of shorebirds, waterfowl, and pelicans which once depended on its waters. Its fish, such as the Owens pupfish and Owens tui chub, also disappeared from the region. The lake bed, now exposed to the desert winds, is a modern reminder of how much the desert depends on water.

Many experience the wildlife of the Mohave only in their headlights as they speed up Highway 395, catching a ghostlike glimpse of a passing kangaroo rat. The wildflower extravaganza blanketing every square millimeter of desert floor in spring is spectacular but brief, and missed by most. The assumption that there is no life in the desert is all too common, and leads to the misconception that the desert can be abused. Many people mistakenly equate the desert's inhospitable environment with an indestructible one. The future for much of the Mohave is in question; its fate rests with those who can see past its inhospitable greeting and feel the warmth of its embrace.

Saline Valley. *This is one of the great treasures of the Mohave Desert. Only a couple of valleys away from Death Valley, Saline Valley is a rock-strewn refuge for a huge marsh.*

Mohave Ground Squirrel

In a very unforgiving environment, the threatened Mohave ground squirrel persists and prospers. The strategy this small squirrel employs is unique, especially when compared to other small mammals living in the same habitat.

Like most ground squirrels, the Mohave lives in underground burrows. What is unique about this species it that it spends up to seven months of the year in its burrow in hibernation. In a desert, one would think the time to come up would be in the winter, but the Mohave ground squirrel makes its surface appearance during the heat of the summer, March to August.

The ground squirrels associate with creosote, Joshua tree, or shadscale plant communities on flat or level plains. Here they depend on the fruits and seeds of the desert plants. During their short time above ground, they breed, bear four to six young, and put on enough fat to last seven months underground.

Their highly specialized lifestyle and habitat requirement has made them very susceptible to man's encroachment. As with most endangered species, the loss or degradation of the habitat has caused the extinction of many populations and isolated many more.

The Mohave ground squirrel is now in unprotected limbo. The species was expected to be declared endangered, but in 1992 the state game commission delisted the Mohave ground squirrel for economic reasons. Not only is this illegal, as stated in the California Endangered Species Act, it sets a very dangerous precedent. It is another example of politics and not science deciding the future of our natural heritage.

Death Valley. *One of the Mohave's greatest landmarks is Death Valley, and in the valley everyone should spend some time at Zabriskie Point. This magnificent view overlooking the valley below takes in the inspiring foothills which rise from below sea level.*

GREAT BASIN

...forms the western limit of the fertile and timbered lands along the desert and mountainous region included within the Great Basin-a term which I apply to the intermediate region between the Rocky Mountain and the next range, containing many lakes, with their own system of rivers and creeks and which have no connection with the ocean.

Colonel John C. Fremont, 1844

Labeled the cold desert, the Great Basin is the largest desert in North America, but the smallest in California. Its southwestern corner extends into the state, taking over the creosote plains of the Mohave and replacing them with its own unique calling card, the sage. Called by many the "sea of sage," the Great Basin is as diverse and specialized a desert as California's other two.

Its boundaries follow a line north of the Owens Valley, going south along the eastern Sierra, and then turning east near the town of Olancha. It runs through Death Valley, making up the northern half of the monument. It is what many call the high desert because most of it is above 4,000 feet in elevation.

Climbing Sherwin Grade after leaving Bishop, the Great Basin meets the eye as the sage greets the nose. A visual difference is immediately apparent. Its flat plains are covered with sage, and there are always mountains somewhere in the picture. In the summer, the sage retains a "green" appearance, so the Great Basin Desert doesn't look like the other deserts.

Sagebrush could possibly be the most abundant shrub of the West. In many places it has a death grip on the environment, forming a subcanopy in which other plants attempt to grab hold. Where life-sustaining water is abundant, the sage might grow seven feet tall; where it is not, the sage grows only inches. Historically, sage was not such a dominant plant. Not until grazing eliminated the perennial grasses did the sage gain its hold.

One symbol of the Great Basin and the sagebrush is the sage grouse. These solitary birds, the size of small turkeys, rule the sage. Living their lives (nesting in and foraging on sage) in almost total secrecy, their courtship displays break all the rules. During a short time in late winter, they gather in large numbers out in the open for their

Conway Summit. *Where the Great Basin and the Sierra Nevada collide, the sage meets the aspens in a spectacle of color.*

LEFT: ***Crowley Lake.*** *The male sage grouse struts out on the lek, puffs out his chest, and fills his air sacs only to expel them in a funny-sounding gurgle. This is all part of his ancient breeding ceremony.*

mating rituals. Their strutting grounds, or lek, are ancient meeting grounds where the grouse have been performing the same ceremony since long before recorded time.

The males' sleek appearance gives way to their parade dress. In fine fashion to attract a mate, the males flare their tail fans and strut. This in itself is not enough to attract a mate. The males have large air sacs on their breasts which they inflate, then expel to make a unique noise. The combination of this drives the females wild; once paired up, they mate and then venture back into the sage. After ten days of this yearly gathering, they all disappear into the sea of sage until the next winter.

The sage is home to North America's fastest land mammal, the pronghorn antelope. Classified as a sagebrush plains species, its white rump patch makes it conspicuous in the sea of gray-green. More than likely, that patch is all you will see, since the pronghorn is wary of humans. Once ranging through most of the Great Basin and California grasslands, they were hunted nearly to extinction by the miners in the 1850s. They have made a slow but steady comeback and can now be seen gracing their sage habitat in areas such as the hills behind Bodie.

Many other species reside in the realm of the sage, all specializing in living where the snow is frequent and winter long, where the summer sun bakes the earth's crust solid. Some, such as sage sparrows and sage thrashers, spend all their time scratching a life from the sage. They utilize the sagebrush

Bodie Bluff. *One of the mysterious mammals of the sage is the pika, also known as rock rabbit. Pikas are best known for their high altitude homes in ranges such as the Sierra, but these adventurous animals also inhabit many of the mountain islands in the sea of sage. Currently, studies are trying to understand how such specialized creatures can be so widespread.*

Great Basin. *The smallest bird of prey, though not officially considered one, is the loggerhead shrike. That specialized bill with its deadly hook seizes prey easily and tears it up. A look at one of its victims (next picture) shows why the shrike is also known as the butcherbird.*

Great Basin. *Named for the plant it inhabits, the sage thrasher is a noisy bird. This particular bird made its presence known to a loggerhead shrike, which has a distasteful habit of impaling its victims on spikes, in this case a barbed wire fence.*

as a home for their nests, constructing them out of its shredded bark. Others, such as the least or sagebrush chipmunk, adapt by going underground during the cold of winter, hibernating until the warmth comes to the basin. And the pygmy rabbit, the smallest of all rabbits, digs a burrow under the sage and is believed to consume nothing but sage.

The wealth of the Great Basin is readily apparent to the visitor venturing here with an eye for progress. More than the Colorado or Mohave, the Great Basin has already been exploited. The pronghorn's numbers are greatly reduced from historic accounts. The sage grouse, once so very common, is now rare and some believe it is vulnerable to upcoming pressures on its habitat. Discussing the future for any one of the three deserts comes down to the same bottom line. The hard shell they appear to throw up in their own defense really only hides just how vulnerable they are. The sands of time will tell if we are wise enough to have learned the lessons of time, and can avoid leaving our "wagon rut" for others to find in 130 years.

Great Basin. *In the distance the White Mountains rise to interrupt the great sea of sage. Though only a small fraction of the Great Basin lies within California's borders, it is a massive desert.*

SALT MARSHES

Cradles of Life

Miles of (the Franciscan tidelands) barely above the level of the slow-moving water, spread a magic carpet of blending crimsons, purples, and bronzes. Under the creeping mists and subject to the changes of the water, beaten to gold and copper under the sun, it redeems the flat lines of the landscape with the touch of splendor.

Mary Austin, 1903

The U.S. Fish and Wildlife Service recognizes over one hundred types of wetlands or marshes, but none are as well known or spectacular as California's coastal marshes. These marshes and tidal mudflats support a greater variety and number of species and produce more organic material per acre than any other habitat in the world.

Historically they were thought of as "practically worthless." Filling them in has long been sponsored by the government. This has meant a minimum loss of seventy-five percent of California's historic 382,000 acres of salt marsh. Activities such as dredging, draining, filling, plus degradation from pollution, have left areas such as southern California with less than ten percent of its historic salt marsh.

The broad definition of a salt marsh covers most of California's coastal wetlands. A salt marsh is a flat area that receives at least one tide each day. Most marshes receive tidal action twice every twenty-five hours, which is vital for a healthy marsh ecosystem. The saltwater tidal action delivers and removes nutrients that the marsh requires. Many salt marshes also receive freshwater influence

LEFT: ***China Camp State Park.*** *In San Francisco's northern bay, this is a rare locale where the oaks meet the sea. This safe harbor is very important to a number of species, especially the black rail, a threatened species.*

from streams and rivers, bringing needed silt to create new marshes and nutrients to feed the old.

The daily drama of the salt marsh is disappearing from California's landscape. The battle to save the remaining salt marshes and their inhabitants is an ongoing process; the scope of what is at risk can be seen by exploring just a couple of California's coastal wetlands.

Palo Alto Baylands

San Francisco Bay is one of the grandest marshes on the Pacific coast. Its complexity fills volumes of scientific journals, covering everything from the influx of silt from the San Joaquin and Sacramento rivers to the little creatures that make their home at the bottom of the bay. The bay itself is a living organism; spotlighting just the salt marsh does not do it justice. But more than any other ecotope of the bay, the salt marsh shows how endangered an ecosystem it really is.

San Francisco Bay is divided into two sections, north bay and south bay. They both have salt marshes, but they have slight differences in structure that affect the inhabitants they support. The historic marshes of San Francisco Bay once covered 315 square miles; they are now reduced to less than sixty square miles. Tucked away in a corner of the south bay, however, is the slice of San Francisco Bay salt marsh that best represents the bay, Palo Alto Baylands.

Palo Alto Baylands is a 1,700-acre preserve and a prime example of a working tidal marsh. Though not perfect, since some of its prime edges have been lost, the ebb and flow of life can be seen here, telling the marsh's story completely. There are a number of key players that make this a working marsh—the tides, the plants, and those who inhabit the marsh.

The mudflats of the marsh have long been thought of as wasteland, as was the marsh itself. But the mud supports thousands of inhabitants not visible from the surface. Oysters, clams, amphipods, isopod crustaceans, and many kinds of worms filter feed as the waters of the marsh flow by. Some only come out when the tide covers their muddy homes, disappearing into its slimy depths once the tide goes out. They are fed by the rich zooplankton that supports much of the life in the bay.

These invisible creatures are crucial food for the hundreds of thousands of shorebirds depending on the marsh. In the waters and mudflats of the baylands, thirty or more bird species can be identified in just minutes. Shorebirds big and small, from marbled godwits to western sandpipers, feed shoulder to shoulder. They are able to do this because they have different bill lengths; each searches at a certain depth in the mud, harvesting the little creatures within reach of its bill.

Ducks take advantage of the food source at high or low tide. The northern pintail and American wigeon typically stick their tails in the air and their heads down to sift for

Palo Alto Baylands. *The setting sun and the mood it casts over the waters of the baylands hides the fact that there is an airport and city in view. For those who venture to the baylands, these blinders are set in place by the wonders one sees while experiencing its marvels and is what brings so many back again and again.*

Palo Alto Baylands. *The salt marsh harvest mouse lives and thrives within the confines of the pickleweed of the salt marsh. It eats the pickleweed as if it were sweet corn and scrambles through it as if it were a paved highway. An endangered species, its future is in grave doubt; no one knows how many of these mice still survive.*

small prey at the water's bottom. Other ducks, such as the green-winged teal, take advantage of low tide to sift the small puddles of water for food. During different times of the day these ducks will create rafts of thousands, all with their heads tucked into their wings.

The only fresh water a salt marsh receives is from an in-flowing creek or rain, so the plants that make up the marsh flora are adapted to drinking salt water. Their ability to survive heavy to light doses of salt water determines where plants can exist in the marsh. Since cordgrass can take complete emergence, it normally grows on the very outer edge of the marsh, at the low tide zone.

The pickleweed can tolerate occasional submergence with periods of drying out, so it is the next plant up from the water's edge, at the high tide zone. The pickleweed is a dominant plant in most marshes, spreading in large carpets and supporting much of the life of the marsh. Its name comes from its appearance; pickleweed looks like pickles strung together. The "pickles" are actually its leaves, which turn bright red in the fall and winter before falling off and enriching the mud of the marsh.

One of the most amazing inhabitants of this marsh is the salt marsh harvest mouse, an endangered species. There are actually two subspecies inhabiting San Francisco Bay; the one at Palo Alto Baylands is the southern subspecies. This mouse has adapted to life in a salt-dominated environment; it is dependent on the pickleweed as its food source and is able to drink salt water. Smaller than the average house mouse, the salt marsh harvest mouse reproduces in large quantities, like a typical mouse. Sometimes it takes advantage of small bird nests constructed in the marsh for its own home.

The mice live most of the year scurrying around the pickleweed, going about the business of survival. However, their world turns upside down during the first winter

California Clapper Rail

The California clapper rail is one of three clapper rails inhabiting California; all are endangered. In the last few years, its population has plummeted from over thirteen hundred individuals to less than three hundred. The loss of habitat is only part of the cause; most of the blame rests with the red fox. This exotic transplant to the California landscape is a master hunter whose only natural enemy is the coyote. The fox came into San Francisco Bay only five years ago, but has since made its presence very well known. It has severely impacted the California clapper rail, salt marsh harvest mouse, and California least tern, plus the shorebirds and ducks that inhabit the marsh.

Part of the problem in helping the clapper rail is the conflict with preserving the salt marsh harvest mouse. Both are endangered species and share the same marsh habitat, but they depend on different levels. The mouse utilizes the pickleweed in the high tide zone while the rail utilizes the cordgrass of the low tide zone. In protecting one species or tidal zone over the other, one of the species might suffer, possibly even perish. Controlling the red fox and making the judgment call as to which plant species or tidal zone is managed will be a difficult decision facing refuge managers in the near future.

For the clapper rail, like many endangered species already stressed from loss of habitat, it only takes one additional pressure, such as the red fox, to push them to extinction. Whatever the outcome, the future for the California clapper rail is in grave doubt.

Palo Alto Baylands. *In the calm waters that provide refuge to so many birds, these American avocets have their minds only on courtship.*

high tides, when they must run to the highest elevation of the marsh to escape the drowning tides. This extreme concentration of mice brings every heron, egret, and other predator out to take advantage of the feast. Historically at the baylands, the egrets and herons would line up shoulder to shoulder. They would arrive just minutes before the tide peaked, wait for it to force the mice from cover, then feast on the vast majority of the mice population. This would last for ten minutes or more, then with the mice gone, the egrets and herons would disperse back into the marsh.

Joining this fray would be another endangered species, the California clapper rail. It also took advantage of any mouse it found trying to wait out the high tide in the marsh. This was the way nature maintained a healthy ecosystem, but today there are no large concentrations of egrets or herons feasting on mice. The mouse population has fallen so dramatically that this event has not been witnessed for over a decade.

It is during this same time that one might be able to catch a glimpse of the threatened California black rail. This sparrow-sized bird is also flushed to the fringes of the marsh by the high tide and can fall prey to the egrets and herons. The black rail population once covered much of the state's freshwater and saltwater marshes, but their numbers have been vastly reduced in the last decades.

What really makes Palo Alto Baylands stand out from any other marsh in the bay is that its wildlife is visible in broad daylight for all to see. During the winter, high tides are the best times to witness the struggle of life that each inhabitant must overcome every minute of the day. The California clapper rail, the ghost of the marsh, can be as close as two feet away as it waits out the high tide. And if you are really lucky, you will catch a glimpse of a salt marsh harvest mouse scurrying over the pickleweed.

If you want to see this, you should act quickly because the marsh is slipping away as you read this. Still being developed, dredged, and polluted, it is by no means safe. Even Palo Alto Baylands is not safe from the many complicated factors slowly killing the marsh. Such things as the influx of fresh water from sewage treatment plants is altering the salinity of the salt water to the detriment of the plants, creating brackish marshes where nothing can live. More than any other habitat in California, destruction of the marsh can be seen on a daily basis; if action is not taken very soon, the marsh will survive only in history books!

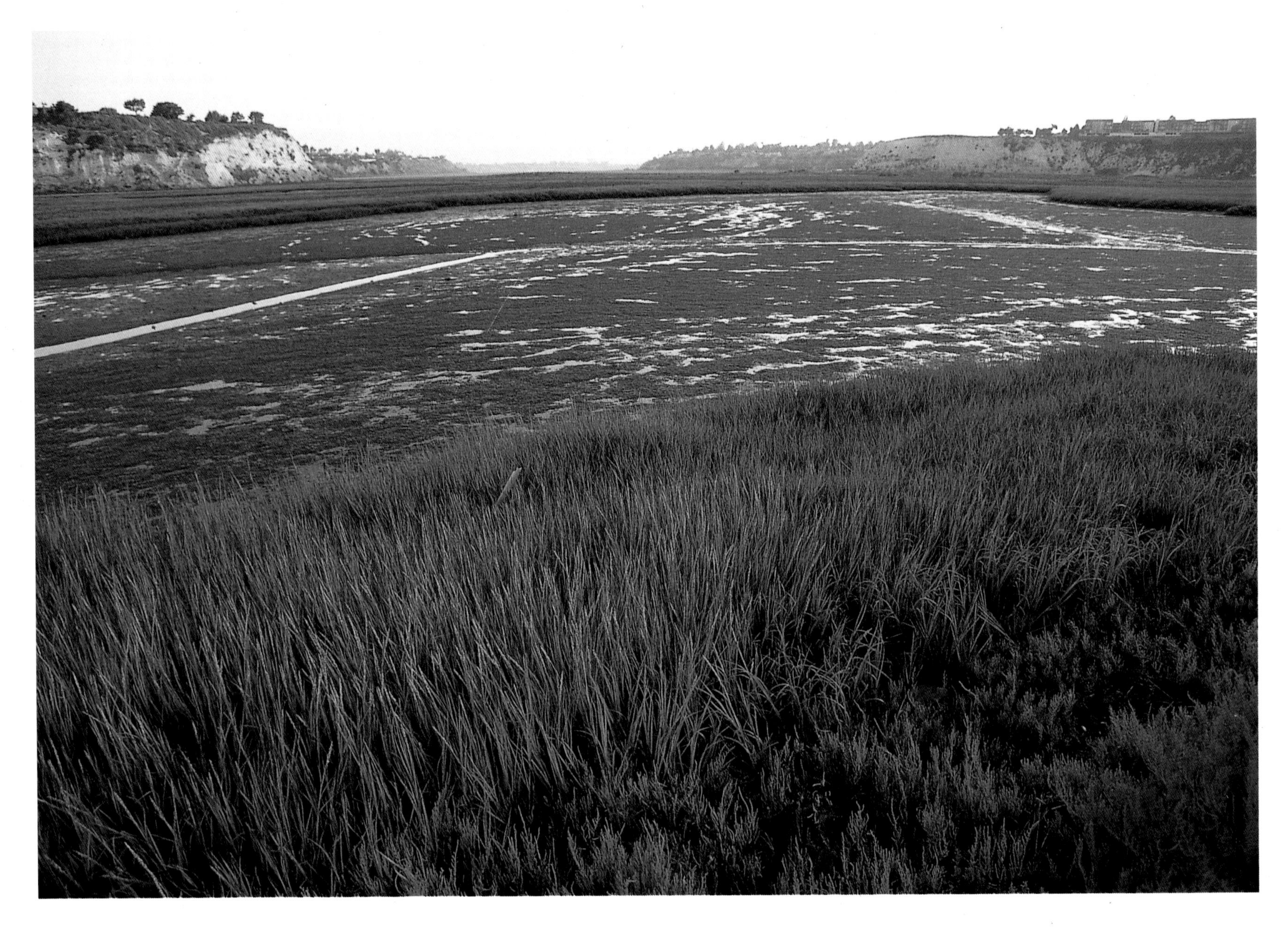

Upper Newport. *On the bluffs surrounding the bay the homes are packed in, side by side. But down below them is one of southern California's last tidal marshes.*

Upper Newport

California was first discovered by explorers who landed in the lagoons of the vast marshes that defined much of the southern coast. These large flat expanses made for natural harbors, and the wealth of wildlife was easy plunder to refill the ships' holds. Sea otters, harbor seals, seabirds, and shorebirds totaling millions strong led these explorers to believe they had found the fabled island of riches, so they named the land California.

In a way they were right, as the new land provided immense wealth. The marshes of the south coast were among the first habitats to be overcome by settlements and exploitation. In the two hundred years to follow, and especially in the last fifty, over ninety percent of those historic marshes vanished. But one of those vast lagoons amidst the tapestry of a salt marsh survived and remains as a window into the past.

Upper Newport, located in Orange County, has managed to maintain its identity through years of tremendous pressure from development. Its 752 acres, encased in white sandstone bluffs, are surrounded on all sides by housing and harbors; its only open space is the sky above.

A true tidal marsh receiving tidal action twice every twenty-five hours, the rich mudflats and vegetation host thirty thousand birds on any given day in August. And when the tide is in and covers the marsh, fish of every description swim in to feed on the rich nutrients the marsh provides.

Upper Newport actually is home to six habitat types: marine, salt marsh, brackish water, freshwater marsh, riparian, and coastal sage scrub. They all thrive within their own niches and together make the salt marsh immensely healthy and productive.

The salt marsh is biologically structured much like its northern cousin. The lower tidal zone is characterized by cordgrass, the higher tidal zone by pickleweed. At those locations where fresh water comes in, a brackish water marsh, typically supporting bullrushes, breaks up the flat landscape. It is the total ecosystem of Upper Newport that supports and attracts 165 bird species to its waters.

The most obvious are the ducks and shorebirds. As with all marshes in California which are part of the Pacific flyway, Upper Newport is a major migration stopover. The hundreds of thousands of birds that depend on Upper Newport use every inch at low and high tide. The dabbler ducks, wigeons, and pintails can constantly be seen with their tails in the air. The feeding teals dot the mudflats at low tide, and create a collage of mixed colors on the water surface at high tide.

The shorebirds that come to Upper Newport fill the dreams of many birders. Flocks of hundreds can pack into a square yard of mud at low tide. The most common are the western sandpipers, dowitchers, dunlins, and godwits. American avocets are year-round residents, swishing their bills through the mudflats looking for a meal.

In the cordgrass and pickleweed are other marsh birds. The most celebrated is the endangered light-footed clapper rail. Upper Newport is home to two-thirds of the world's remaining population of four hundred birds. Their close proximity to hundreds of human visitors every day has made the subpopulation on the west end of Upper Newport very visible. It is easy to watch their entire life biology unfold from only ten feet away. Everything from territorial disputes to mating, family raising, and sunbathing is no more than a glance away.

The clapper rail shares its habitat with the threatened Belding's savannah sparrow. This subspecies has thirteen other cousins in California, but it is the only one in danger of extinction. This somewhat dark sparrow thrives in a few marshes, but has lost so much of its historic marsh habitat that it is in the process of being upgraded to endangered.

In these same pickleweed and cordgrass

Upper Newport. *Dressed in their winter plumage, this flock of American avocets huddles together for an afternoon siesta.*

communities live small mammals that few ever see. The salt marsh wandering shrew and southern California salt marsh shrew are both candidates for listing. These two- to three-inch mammals and their cousins around the state have been mostly unnoticed, yet many are vanishing right before our eyes. Their very secretive life has hindered efforts to understand and protect them. With marsh habitat vanishing, their extinction will be an end result.

Upper Newport has two very important man-made islands, which were created to benefit another endangered species, the California least tern. This small, migratory species is one of the success stories of endangered species, thanks to the efforts of very dedicated individuals. Dependent on the sands adjacent to rivers where they meet the ocean, the terns have been systematically pushed out of their ancestral homes by sunbathers and development. These tenacious little birds have hung on. With the creation of man-made islands at Upper Newport and other southland locales, and the use of wooden decoy terns to attract them, they have colonized secure new locales. Their numbers are slowly but steadily increasing, a very encouraging activity in the world of endangered species.

These same islands have also benefited other bird species. The snowy plover, a candidate for listing, is one of those species. Like the tern, it has lost much of its nesting

LIGHT-FOOTED CLAPPER RAIL

The light-footed clapper rail has lost as much as ninety percent of its historic habitat, but it still has a fighting chance of survival.

The clapper rail's story is understood best by the experience with the red fox at Anaheim Bay, a wildlife refuge. In the eighties, the rail population crashed to ten individual birds at the refuge. The cause was red fox predation. After more than two hundred red fox were removed from the refuge, the clapper rail population rebounded to one hundred individuals.

At Upper Newport, development threatens to cut off the main path the coyote utilizes to enter the preserve. The red fox has yet to take hold at Upper Newport because the coyote keeps the fox at bay, but if the coyote is eliminated from the preserve, the clapper rail, least tern, savannah sparrow, and other bird species will all be pushed to the very edge of existence by the red fox. For the clapper rail, Upper Newport is its major refuge. The rail's future depends on our actions and understanding of the entire ecosystem and the dependencies of those elements not directly housed inside the white sandstone bluffs.

habitat on the sandy beaches. New colonizers, the elegant tern and black skimmer, have joined these birds on the islands. These species of special concern, along with all the other birds, pack the islands to the edges. But the bay is healthy enough to provide for all these migratory birds, so they can successfully raise their young.

The habitats that fringe the bay are also important. The peregrine falcon, gone from the bay for over a decade, has now returned and is believed to nest on the sandstone bluffs. This endangered species is also a limited success story that now depends on the bay. The California gnatcatcher is another of the species supported by the coastal sage scrub of Upper Newport.

The wealth of Upper Newport can easily be measured by the number of endangered species that not only survive in its waters, but manage to thrive. Its uniqueness and the importance of maintaining it as a preserve are also apparent. Development still has its sights on Upper Newport; its safety is far from secure. Pollution flowing into the marsh via the freshwater streams, and development of corridors that prevent the coyote from getting into the marsh are of major concern. A major spill or elimination of the coyote could spell extinction for this vital marsh, a loss that could never be reversed in our lifetimes.

Upper Newport. *In the calm water of low tide, this great egret searches out one last meal before dark.*

Above: ***Upper Newport.*** *Though they are free of the nest not long after hatching, the young least terns are still dependent on their parents to bring them food. But there are those moments when a warm breast and a little shade are just as welcome as a cold, slimy fish.*

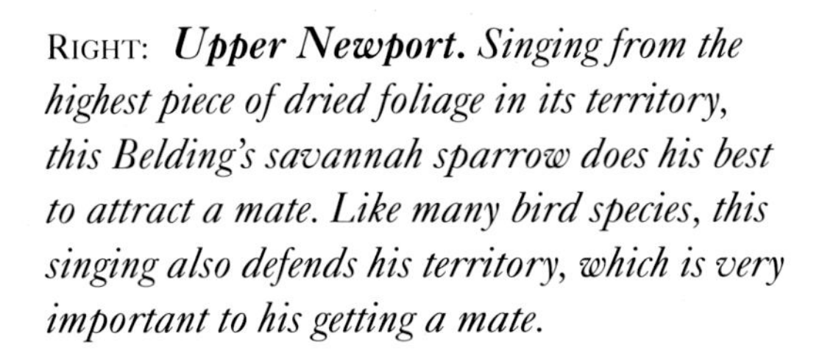

Right: ***Upper Newport.*** *Singing from the highest piece of dried foliage in its territory, this Belding's savannah sparrow does his best to attract a mate. Like many bird species, this singing also defends his territory, which is very important to his getting a mate.*

AFTERWORD

Within its borders, California has the greatest biodiversity of any region in North America. After more than a century of agriculture, development, housing, and recreation, this diversity is a testament to the grandeur and wealth of the Golden State.

This book shows only a sample of the heritage that belongs to us all. My hope is that it has opened your eyes, minds, and hearts to the spectacle of wildlife and wild places that still exist in California--right outside our doors--to be explored, cherished, and preserved.

In a hundred years, will a photographer be able to celebrate California's landscape in photographs? Will we as citizens be able to put aside our personal ambitions to ensure that we preserve this legacy to pass on to future generations? Can we recover from the destructive, shortsighted attitudes of past and present?

Some readers will consider this book a call to activism, others might see it as a pessimistic look at the future. If only these impressions remain of the book, then I will have partially succeeded. If, ten years from now, this book stands only as a testament to the beauty that once graced California's land, then we all will have failed.

The point of this book is not to depress people, nor to paint a picture of hopelessness. We could list the specific people causing the loss of our natural heritage. Developers, politicians, conservationists, and especially the public would have to be included on that list, but pointing fingers to place blame only hurts the habitat and wildlife that is at risk.

In the process of listing a species as endangered, its habitat is listed as critical to its survival. This is backwards. The habitat should be listed as endangered, with the species as a critical inhabitant. The battle over California's remaining wild landscape will make front page headlines often in the coming years. More listings for endangered species will be fought; the fate of specific habitats will be decided in courts, not in biological round tables.

The biological welfare of California's habitats and wildlife is governed by politicians, not scientists. Effecting a positive change requires an awareness of the magnificent heritage we are at risk of losing. Take a stance in protecting the wildlife in your own backyard; care about the wild legacy you are leaving to your children. All of us doing this together will make an overwhelming difference.

The process will not be easy or painless. Through careful scientific study, soul-searching, and compromise from developers and conservationists, it is realistic to preserve California's natural heritage. Some developers might lose precious profits, and conservationists might lose some precious species and habitats. But without a unified effort, time will pass and our landscape and wildlife will be the losers.

The antics of San Joaquin kit fox pups playing outside their den entrance, the busy activities of the giant kangaroo rat as it goes about its nightly chores,the ruckus of the clapper rail defending its territory, the salt marsh harvest mouse eating pickleweed as if it were sweet corn—all these are daily events still played out on the California stage. We cannot turn back the clock and see again the grizzly bear or Santa Barbara song sparrow, but we owe future generations the opportunity to know the wildlife and habitats we have been privileged to enjoy.

Why should we care so much about endangered species and habitats? This is the question which we all come to at some point. The health of the environment, on which we as a society depend, can be likened to a huge puzzle. And like any puzzle, every piece must be in place in order to see the whole picture. With each species and habitat that we lose, we also lose a piece of the picture forever.

To learn more or find out how you can help preserve California's wild heritage write to:

California Department of Fish & Game
Natural Heritage Division
1416 Ninth Street
Sacramento, CA 95814

California Native Plant Society
909 12th Street
Sacramento, CA 95814

California Wilderness Coalition
2655 Portage Bay East, Suite 5
Davis, CA 95616

Defenders of Wildlife
5604 Rosedale Way
Sacramento, CA 95822

Environmental Protection Agency
215 Fremont Street
San Francisco, CA 94105

Friends of the River
909 12th Street
Sacramento, CA 95814

Friends of the Sea Otter
P. O. Box 221220
Carmel, CA 93922

National Audubon Society
Western Regional Office
555 Audubon Place
Sacramento, CA 95822

National Marine Sanctuaries Program
1825 Connecticut Ave NW
Washington, DC 20235

National Wildlife Federation
1412 16th Street NW
Washington, DC 20009

The Nature Conservancy
785 Market Street
San Francisco, CA 94103

The Sierra Club
730 Polk Street
San Francisco, CA 94109

USDA Forest Service
630 Sansome Street, Room 559
San Francisco, CA 94111

U.S. Department of the Interior
Bureau of Land Management
2800 Cottage Way, Room E-1803
Sacramento, CA 95825

U. S. Department of the Interior
Fish & Wildlife Service
2800 Cottage Way, Room E-1803
Sacramento, CA 95825

The Wilderness Society
1400 I Street NW
Washington, DC 20005

World Wildlife Fund
1255 23rd Street NW, Room 200
Washington, DC 20037

BIBLIOGRAPHY

REFERENCES QUOTED:

Pages 114, 127
Austin, Mary. *The Land of Little Rain.* 1903. Reprint. University of Austin Press, 1974.

Pages 56, 60, 85, 105
Farquhar, Francis P., ed. *Up and Down California in 1860-1864, The Journal of William H. Brewer*, Yale University Press, 1930.

Pages 13, 31, 34, 65, 103
Hittell, Theodore H. *Adventures of James Capen Adams Mountaineer and Grizzly Bear Hunter of California*, Scribner's, 1912.

Page 15
Holing, Dwight. *California Wild Lands*, The Nature Conservancy, 1988.

Page 77
Marryat, Frank. *Mountains and Molehills*, Harper & Brothers, 1855.

Pages 59, 120
Vincent, Stephen, ed. *O'California,* Bedford Arts, 1990.

OTHER REFERENCES:

Atwood, Jonathan L. *Status Review of the California Gnatcatcher*, Manomet Bird Observatory, 1990.

Bakker, Elna S. *An Island Called California.* 2d edition. University of California Press, 1984.

Bent, Arthur Cleveland. *Life Histories of North American Birds of Prey, Part One*, Dover Publications, 1961 edition.

———. *Life Histories of North American Birds of Prey, Part Two,* Dover Publications, 1961 edition.

———. *Life Histories of North American Cuckoos, Goatsuckers, Hummingbirds, and Their Allies*, Dover Publications, 1989 edition.

———. *Life Histories of North American Flycatchers, Larks, Swallows, and Their Allies*, Dover Publications, 1963 edition.

———. *Life Histories of North American Wagtails, Shrikes, Vireos, and Their Allies*, Dover Publications, 1965 edition.

Braun, Suzanne E. "Home Range and Activity Patterns of the Giant Kangaroo Rat," *Journal of Mammalogy* 66(1): 1-12 (1985).

Burt, William Henry, and Richard Philip Grossenheider, *A Field Guide to the Mammals of America North of Mexico*, Houghton Mifflin Co., 3d edition, 1976.

Culbertson, A. E. "Observations on the Natural History of the Fresno Kangaroo Rat," *Journal of Mammalogy* 27(3): 189-203 (1946).

Forbes, Alexander Esq. *California: A History of Upper and Lower California*, Smith, Elder & Co., 1839.

Garrett, Kimball, and Jon Dunn. *Birds of Southern California*, Los Angeles Audubon Society, 1981.

Grinnell, Joseph. "Habitat Relations of the Giant Kangaroo Rat," *Journal of Mammalogy* 13(4): 305-320 (1932).

Grinnell, Joseph, and A. H. Miller. *The Distribution of the Birds of California*, Cooper Ornithological Club, 1944. Reprint. Artemisia Press, 1986.

Harris, John H. *Mammals of the Mono Lake-Tioga Pass Region*, David Gaines/Kutsavi Books, 1982.

Hutchings, J. M. *Scenes of Wonder and Curiosity in California*, Hutchings, 1862.

Ingles, Lloyd G. *Mammals of the Pacific States*, Stanford University Press, 1965 edition.

Jameson, E. W. Jr., and H. J. Peeters. *California Mammals*, University of California Press, 1988.

Johnsgard, Paul A. *North American Owls: Biology and Natural History*, Smithsonian Institution Press, 1988.

Jones & Stokes Associates. *Sliding Toward Extinction: The State of California's Natural Heritage*, The California Nature Conservancy, 1987.

Kreissman, Bern. *California: An Environmental Atlas & Guide*, Bear Klaw Press, 1991.

Lowe, David W., J. R. Matthews, and C. J. Moseley, eds. *The Official World Wildlife Fund Guide to Endangered Species of North America*, 3 vols., Beacham Publishing, 1990-1992.

Matthiessen, Peter. *Wildlife in America*, Viking Press,1987 edition.

Monson, Gale, and Lowell Sumner, eds. *The Desert Bighorn*, University of Arizona Press, 1980.

Niehaus, Theodore F., and C. L. Ripper. *A Field Guide to Pacific States Wildflowers*, Houghton Mifflin Co., 1976.

The Wildlife Society. *California Gnatcatcher Workshop*, The Wildlife Society, Southern California Chapter, 1991.

California Department of Fish and Game Publications:

Airola, Daniel A. *Guide to the California Wildlife Habitat Relationships System*, Department of Fish and Game, 1988.

Bleich, Vernon C. *Amargosa Vole Study*, Job Final Report W-54 R-10, Department of Fish and Game, 1980.

Draft Five Year Status Report: Amargosa Vole, Department of Fish and Game, 1989.

Five Year Status Report, various species, Department of Fish and Game, 1987-1988.

Knapp, Donna Kay. *The Fresno Kangaroo Rat Study*, Department of Fish and Game, 1975.

Laymon, Stephen A., B. A. Garrison, and J. M. Humphrey. *Historic and Current Status of the Bank Swallow in California*, Department of Fish and Game, 1987.

Littlefield, Carroll D. *The Status and Distribution of Greater Sandhill Cranes in California, 1981*, Department of Fish and Game, 1982.

Massey, Barbara W. *A Census of the Breeding Population of the Belding's Savannah Sparrow in California*, Department of Fish and Game, 1977.

Mayer, Kenneth E., and W. F. Laudenslayer Jr., eds. *A Guide to Wildlife Habitats of California*, Department of Fish and Game, 1988.

Moyle, Peter B., J. E. Williams, and E. D. Wikramanayake. *Fish Species of Special Concern of California*, Department of Fish and Game, 1989.

1990 Annual Report on the Status of California's State Listed Threatened and Endangered Plants and Animals, Department of Fish and Game, 1991.

1989 Annual Report on the Status of California's State Listed Threatened and Endangered Plants and Animals, Department of Fish and Game, 1990.

1988 Annual Report on the Status of California's State Listed Threatened and Endangered Plants and Animals, Department of Fish and Game, 1989.

O'Farrell, Michael J., and C. Uptain. *Assessment of Population and Habitat Status of the Stephens' Kangaroo Rat*, Department of Fish and Game, 1989.

Steinhart, Peter. *California's Wild Heritage: threatened and endangered animals in the golden state*, California Department of Fish and Game, 1990.

Williams, Daniel F. *Distribution and Population Status of the San Joaquin Antelope Squirrel and Giant Kangaroo Rat*, Department of Fish and Game, 1987.

———. *Mammalian Species of Special Concern in California*, California Department of Fish and Game Report 86-1, 1986.

Williams, Daniel F., and G. E. Basey. *Population Status of the Riparian Brush Rabbit,* Department of Fish and Game, 1986.

Zeiner, David C., W. F. Laudenslayer, Jr., K. E. Mayer, and M. White, eds. *California's Wildlife.* 3 vols. Department of Fish and Game, 1988-1990.

U.S. Department of the Interior, Fish and Wildlife Service Publications:

Anderson, Stanley H. *Yuma Clapper Rail Recovery Plan,* U.S. Fish and Wildlife Service, 1983.

Draft Recovery Plan for the Fresno Kangaroo Rat, U.S. Fish and Wildlife Service, no date.

Faber, Phyllis M., E. Keller, A. Sands, and B. M. Massey. *The Ecology of Riparian Habitats of the Southern California Coastal Region: A Community Profile*, Biological Report 85(7.27), U.S. Fish and Wildlife Service, 1989.

Franzreb, Kathleen E. *Ecology and Conservation of the Endangered Least Bell's Vireo,* Biological Report 89(1), U.S. Fish and Wildlife Service, 1989.

Herbold, Bruce, and P. B. Moyle. *The Ecology of the Sacramento-San Joaquin Delta: A Community Profile*, Biological Report 85(7.22), U.S. Fish and Wildlife Service, 1989.

Josselyn, Michael. *The Ecology of the San Francisco Bay Tidal Marshes: A Community Profile*, FWS/OBS–83/23, U.S. Fish and Wildlife Service, 1983.

Marshall, David B. *Status of the Marbled Murrelet in North America: with special emphasis on populations in California, Oregon and Washington*, Biological Report 88(30), U.S. Fish and Wildlife Service, 1988.

Monk, Geoff. *California Peregrine Falcon Reproductive Outcome and Management Effort in 1981*, U.S. Fish and Wildlife Service, 1982.

O'Farrell, Thomas P. *San Joaquin Kit Fox Recovery Plan*, U.S. Fish and Wildlife Service, 1983.

Report To Congress: Endangered and Threatened Species Recovery Program, U.S. Fish and Wildlife Service, 1990.

Roest, Aryan I. *Morro Bay Kangaroo Rat Recovery Plan*, U.S. Fish and Wildlife Service, 1982.

Salata, Larry. *A Status Review of the California Gnatcatcher*, U.S. Fish and Wildlife Service, 1991.

Shellhammer, Howard S., T. E. Harvey, M. D. Knudsen, and P. Sorensen. *Salt Marsh Harvest Mouse and California Clapper Rail Recovery Plan*, U.S. Fish and Wildlife Service, 1984.

Wilbur, Sanford R., R. M. Jurek, R. Hein, and C. T. Collins. *Light-footed Clapper Rail Recovery Plan*, U.S. Fish and Wildlife Service, 1979.

Williams, Daniel F. *A Review of the Population Status of the Tipton Kangaroo Rat,* U.S. Fish and Wildlife Service, 1985.

Woodard, Donald W. *American Peregrine Falcon*, U.S. Fish and Wildlife Service, 1980.